Landscape Heritage of the World with Google Earth

Andrew Goudie

Landscape Heritage of the World with Google Earth

Springer

Andrew Goudie
School of Geography
University of Oxford
Oxford, UK

ISBN 978-3-031-95554-9 ISBN 978-3-031-95555-6 (eBook)
https://doi.org/10.1007/978-3-031-95555-6

This Springer imprint is published by the registered company Springer Nature Switzerland AG
The registered company address is: Gewerbestrasse 11, 6330 Cham, Switzerland

Acknowledgements Although myself a geomorphologist and landscape Geographer, over the years I have been fortunate to work with, or to meet up with, various Archaeologists. I particularly wish to acknowledge Raymond and Bridget Allchin in South Asia, K. T. M. Hegde, D. P. Agrawal, R. S. Pappu, Cameron Petrie, and S. N. Rajaguru in India, J.-F. Jarrige, Peter Arbuthnot, M. A. Halim in Pakistan, Larry Barham in Zambia, David Price Williams in Eswatini, Qinglin Guo, Wang Xudong, He Qing Huang, Wenwu Chen, Yanwu Wang, and Hong Zhang in China, and Derek Kennet and Adrian Parker in Oman and the UAE. I have also gained more than they may realise from my archaeological colleagues in St. Cross College, Oxford, including Dan Hicks, Julia Lee Thorp, Derek Roe, Chris Gosden, Robert Hedges, Mark Robinson, Amy Bogaard, and Helena Hamerow. As a sixth former I was taken to sites in Wiltshire and Gloucestershire by C. S. Neill, and then went on a school trip to Greece and Italy taking in such sites as Athens, Rome, Olympia, Delphi, Mycenae, and Epidaurus. In my first long vacation, in happier times, I went by, taxi and train to Aleppo, Baalbek, Byblos, and Tyre. In my second long vacation I drove to Greece and back with Robert Wickham, Jim McLeod, and Paul Hopper. As a graduate student I went with Mike Wood to North Africa and visited Cyrene, Sabratha, Leptis Magna, Apollonia, and Volubilis. In Egypt, Nabil Embabi and M. Ashour took me to the Pyramids and the Western Desert, and Stephen Stokes took me to Luxor, Karnak, and Kharga. I have also had a series of family trips with Heather, Amy and Alice, to Zimbabwe, Jordan, Sicily, Istanbul, Egypt, Sardinia, Morocco, Tunisia, Oman, and Crete. I have also learnt about some ancient landscapes from some Historical Geographers, including H. C. Darby, Harold Fox, and Roy Millward at Cambridge, and Frank Emery, Paul Coones, John Patten, John Wilkinson, and Michael Williams in Oxford. Dick Grove demonstrated to me the value of air photo mosaics in landscape research during our circumnavigation of the Kalahari in 1967. As a callow young don I recall being most affably and tolerantly treated by Dame Kathleen Kenyon at St Hugh's College and by Sir Mortimer Wheeler in Cambridge. I am fortunate to live in a modest cottage between two castles on the Isle of Portland, itself located in the county of Dorset, one of the richest locations for heritage sites in the whole of the British Isles. I am grateful to Yanwu Wang, Alice Goudie, and Heather Viles for some images, and I owe a particular debt to Google Earth and Google Scholar for their unrivalled gratis materials. I am very grateful to Robert Doe of Springer and to the production team of Parimel Azhagan, Arumugam Deivasigamani and Sonam Yadav.

Competing Interests The author has no competing interests to declare that are relevant to the content of this manuscript.

Contents

About the Author

Andrew Goudie 2025

Andrew Goudie Ph.D., D.Sc., Emeritus Professor of Geography at the University of Oxford, former Pro-Vice-Chancellor, former Director of the China Centre, Honorary Fellow of Hertford College and the former Master of St. Cross College, is a geomorphologist. He is a recipient of a Royal Medal from the Royal Geographical Society, the Mungo Park Medal of the Royal Scottish Geographical Society, and the Farouk El-Baz Award of the Geological Society of America. He is a Fellow of the British Society of Geomorphologists. He has been Chair of the British Geomorphological Research Group, President of the Geographical Association, and President of the International Association of Geomorphologists. He has written extensively on geomorphology in some 50 books and 300 scientific papers. He is the author of amongst others *The Landforms of England and Wales* (1990), *Great Warm Deserts of the World* (2002), *Arid and Semi-Arid Geomorphology* (2012), *Great Desert Explorers* (2016), *The Human Impact* (2018), *Landscapes and Landforms of England and Wales* (with Piotr Migoń) (2020), *Camel Tracks* (2021), and (with Denys Brunsden), *Geological Pioneers of the Jurassic Coast* (2023). Recently he has used Google Earth in three companion volumes: *Desert Landscapes of the World with Google Earth* (2022) *Landscapes of the Anthropocene with Google Earth* (2023), *and Rocky Landscapes of the World with Google Earth* (2024).

1 Introduction

Abstract

This chapter starts with an introduction to Google Earth for the study of landscapes, and in particular for their morphology, distribution, and changes. It also considers the more general role of air photography and remote sensing in archaeological studies. Google Earth has been immensely useful in the study of World Heritage Sites and in the study of the destruction of sites in recent years, particularly in the Middle East. Examples are given from Syria and Iraq.

Keywords

Google Earth · World heritage · Syria · Iraq · Remote sensing · Air photography

1.1 Google Earth

The rationale of this book is to provide carefully selected images from Google Earth, to discuss what they show, and to provide references to the sites portrayed so that readers can go deeper if they wish. Google Earth (GE) is remarkably useful for examining and appreciating phenomena on the face of the Earth. This book aims to demonstrate its value for understanding heritage landscapes and for appreciating their aesthetics. Many GE images are extraordinarily beautiful, and this volume is, in a sense, a contribution to Geoasthetics. I have deliberately not tried to be fully comprehensive, and so some iconic sites, which are familiar to all, such as the Colosseum in Rome, Pompeii, Stonehenge, and the Acropolis in Athens, have not been included. Equally, I have not attempted to include 'vertical' sites (e.g. the Bamian Buddhas in Afghanistan) because their geometry makes them difficult/impossible to display on GE, and the same applies to some narrow horizontal sites (e.g. ancient tracks).

The book is organised around different types of landscape and the processes that created them, rather than regionally or on the basis of age. It is very apparent that many sites have a long history and may have been modified on many occasions, changing, for example, from temple, to mosque, to church.

GE was launched in 2005 and maps the Earth by superimposing satellite images, air photographs, and GIS data onto a 3D globe, allowing its users to see landscapes from a range of angles (Liang et al. 2018). It is regularly updated, though temporal coverage is inconsistent (Lindsay and Mkrtchyan 2023), and in its basic form is free for users. It is easily accessed via personal computers. Imagery resolution ranges from 15 cm to 15 m. It provides a distance measuring tool, and it is also possible to obtain elevation cross profiles. GE provides near global coverage, so that it is immensely important for determining global distributions of particular phenomena (Goudie 2022; Traganos et al. 2018), including miscellaneous desert landforms (Goudie 2022), Anthropocene features (Goudie 2023), and rocky landscapes (Goudie 2025). However, in general, the resolution is not adequate for the identification of many small landscape features, so that most of the features illustrated in this book are fairly substantial in size. There has also been a tendency to select many images from drylands, as it is in these environments that structures and forms are less obscured by vegetation.

Although GE is much used by archaeologists, it is not always the only or best platform to employ. This is brought out when one considers the Iron Age Hill Fort and Roman Fort (Fig. 1.1) at Hod Hill in Dorset, England, described in the Royal Commission on Historical Monuments (1970, pp. 263–265). Alternative remote sensing resources, including CORONA and Sentinel imagery, are discussed by

A. Goudie, *Landscape Heritage of the World with Google Earth*, https://doi.org/10.1007/978-3-031-95555-6_1

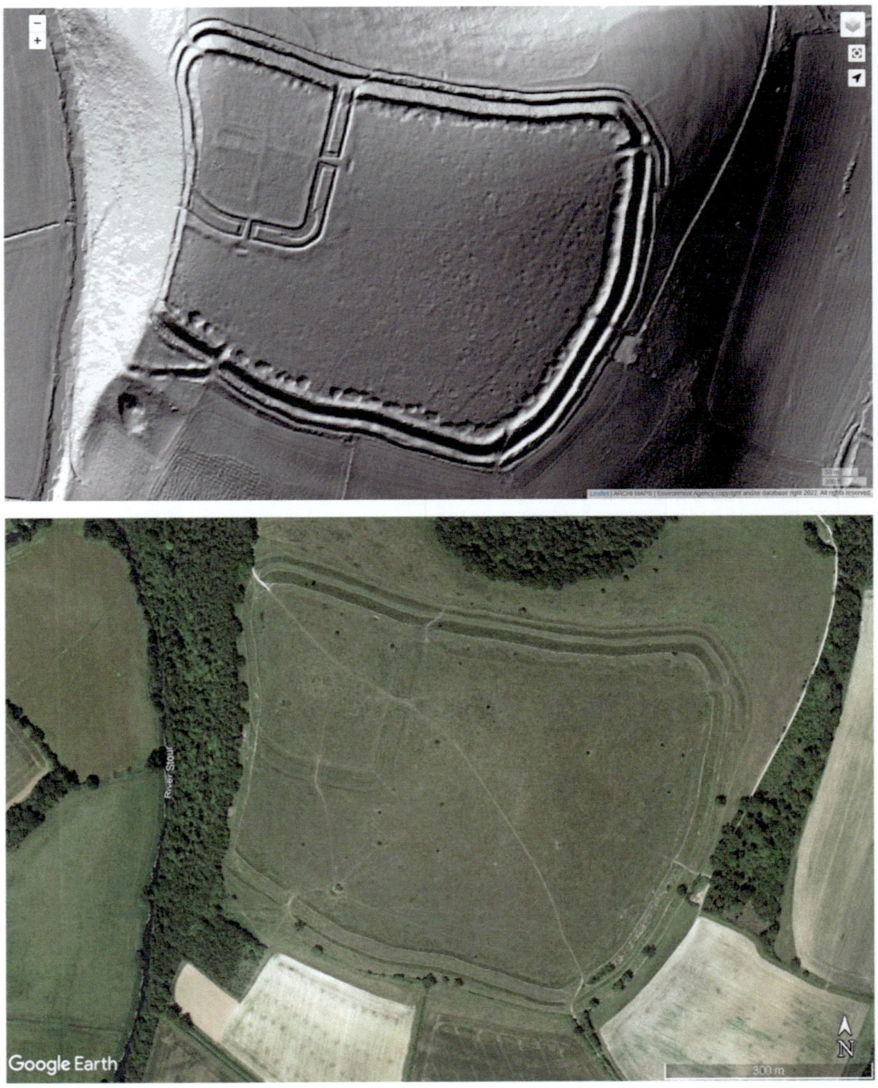

Fig. 1.1 Top: LiDAR image of Hod Hill in Dorset, England. It shows the Iron Age Hill Fort, with a later Roman Fort in the northwest corner. In the east of the image there are large numbers of small circles, which may be remains of huts and of pits. Bottom: Google Earth image of the same location. 50°53′40.92″N, 2°12′16.79″W. Scale bar = 300 m, ©Google Earth

Lindsay and Mkrtchyan (2023), and drone (UAV) images are of increasing value.

Notwithstanding this observation, GE (https://www.google.co.uk/intl/en_uk/earth/) (Accessed 4th April 2025) has been much used in landscape studies, for both teaching (see Lisle 2006; Dolliver 2012; Fisher et al. 2012; Demirci et al. 2013; Tooth 2015; Ortiz-Guerrero and Loizzo 2024) and research (Scheffers et al. 2015). It has also proved to be indispensable for finding new sites (see, for example, O'Reilly and Scott 2015).

In this volume, the latitudes and longitudes of GE images are provided, so that they can be re-located by the reader, as are scale bars. These are located in the bottom right corner of all images. Only vertical images are used in this book, but it is also possible to tilt them.

Two useful texts to provide background on the sites discussed in this book are those by Scarre et al. (2021) and Scarre (2005).

1.2 The Use of Remote Sensing and Air Photography in Archaeology

Remote sensing and aerial photography have both been much used in archaeological studies. They have proved invaluable in finding sites, mapping them, and identifying changes to them through time. One pioneer was the Oxford-trained Geographer O. G. S. Crawford, who after flying with the Royal Flying Corps during World War 1, joined the Ordnance Survey in 1920 as its first Archaeology Officer (Hauser 2008). A landmark publication was *Wessex from the Air* (Crawford and Keiller 1928). Other early aerial archaeologists included Major G. W. G. Allen, who worked primarily in the Oxford area in the 1930s, and D. N. Riley (see Riley 1987). In the Middle East, after World War 1, people like Sir Aurel Stein and P. A. Poidebard pioneered air photography to study archaeological sites and J. H. Breasted undertook air photographic missions over Egypt, the Fertile Crescent and

Iran between 1919 and 1931. After World War 2, J. K. St. Joseph did remarkable work at Cambridge University (e.g. Frere and St. Joseph, 1983). Subsequently there have been various regional studies, including Darvill (1996) in Britain, and Kennedy and others in the Middle East (Kennedy and Riley 1990; Kennedy and Bewley 2004). The University of Cambridge has a huge collection of old air photographs from the UK which show up many archaeological features (https://www.cambridgeairphotos.com/) (Accessed 4th April 2025).

With the coming of space travel, the use of air photography has been supplemented by the use of remote sensing images, including those obtained via GE (Parcak 2009; Kaimaris et al. 2011; Williams et al. 2012). Many archaeologists have used GE (see Luo et al. 2018 for a review) including Kennedy and Bishop in Saudi Arabia (2011), Thomas et al. (2008) in Afghanistan, Sadr and Rodier (2012) in South Africa, and Eidshaug et al. (2024) in Tierra del Fuego. Other examples will be given elsewhere in this book.

1.3 Google Earth and World Heritage Sites

A measure of protection to natural and man-made heritage is provided by World Heritage Site designation (Badman 2010). The 'Convention Concerning the Protection of the World Cultural and Natural Heritage' was adopted by the General Conference of UNESCO on 16 November 1972. As of July 2024, there are a total of 1,223 World Heritage Sites located across 168 countries, of which 952 are cultural, 231 are natural, and 40 are mixed properties.

In this book there are a large number of archaeological sites discussed that have been designated:

Ahom moidams (India), Aleppo (Syria), Anuradhapura (Sri Lanka), Ancient *Ksour* of Ouadane, Chinguetti, Tichitt and Oualata (Mauritania), Angkor Wat (Cambodia), Babylon (Iraq), Bagan (Myanmar), Bali (Indonesia), Banditaccia (Italy), Beaumaris Castle (Wales), Borobudur (Indonesia), Busra (Syria), Cahokia (USA), Cape Coast Castle (Ghana), Caral (Peru), Carthage (Tunisia), Chichĕn Itzá (Mexico), Cidade Velh Turkïye Turkïye a (Cape Verde), Delphi (Greece), Derbent (Russia), Dholavira (India), Dilmun (Bahrain), Dougga (Tunisia), El Djem (Tunisia), El Tajin (Mexico), Epidaurus (Greece), Erbil (Iraq), Fort Jesus (Kenya), Galle (Sri Lanka), Giza pyramids (Egypt), Grand Canal (China), Great Zimbabwe (Zimbabe), Hadrian's Wall (UK), Hagia Sophia (Türkiye), Hatra (Iraq), Hillforts of Rajasthan (India), Hopewell Ceremonial Site (USA), Humberstone (Chile), Iran *Qanat* (Iran), Jesuit Missions of the Guaranis (Brazil and Argentina), Kerkuane (Tunisia), Ksar of Ait-Ben-Haddou (Morocco), Leptis (Lepcis) Magna (Libya), Limes Germanicus (Germany), Lugo (Spain), Machu Picchu (Peru), Maras salt (Peru), Ma'rib (Yemen), Medina Azahara (Spain), Meroë (Sudan), Merv (Turkmenistan), Mohenjo-Daro (Pakistan), Monte Albán (Mexico), Mozambique Island (Mozambique), Mozu Kufungun (Japan), Mycenae (Greece), Nazca Lines (Peru), Nisa (Turkmenistan), Okjeon tumuli (Korea), Olympia (Greece), Paestum (Italy), Paharpur (India), Palmanova (Italy), Palmyra (Syria), Persepolis (Iran), Petra (Jordan) , Pont du Gard (France), Rani Ki Vav (India), Rice terraces (Philippines), Sabratha (Libya), Samarra (Iraq), Segovia Spain), Serpent Mound (USA), Slate landscape of North Wales (Wales), Stonehenge and Avebury (England), Su Nuraxia di Barumini (Italy), Suoyang (China), Syracuse (Italy), Takht-e Soleyman (Azerbijan), Teotihuacan (Mexico), Thebes (Egypt), Timgad (Algeria), Visby (Sweden), Ziggurat of Ur (Iraq).

1.4 Google Earth and Threats to Sites

GE has proved to be important for demonstrating some of the threats that major archaeological sites have been exposed to. These threats include damage by natural events (such as earthquakes, salt decay, landslides, and floods), damage by human infrastructure developments (including urbanisation, pipelines, mining, dam building and road construction), looting of artefacts, and, perhaps most concerning of all, the damage caused by warfare and by ideologically driven site destruction. It has been much used as part of the EAMENA project (Rayne et al. 2017b; Harkel and Bewley 2018) (https://www.classics.ox.ac.uk/eamena-endangered-archaeology-middle-east-and-north-africa) (Accessed 4th April 2025).

Ideologically driven site destruction has been of especial significance in the Middle East over the last two decades, so in this section there follows a discussion of particular sites in Syria and Iraq. In Table 1.1, there is a list of some papers that have addressed this theme.

The first site to be discussed is Apamea in Syria (Fig. 1.2). Located on the right bank of the Orontes River, this was an ancient Greek and Roman city. It is now called Qalaat al-Madiq. It had many impressive ancient remains, including the Great Colonnade, which ran for nearly 2 km, making it amongst the longest in the Roman Empire, and the Theatre, one of the largest surviving theatres of in the Roman

Table 1.1 Papers on the use of remote sensing, including GE, to detect and demonstrate Middle Eastern site destruction

Paper	Location
Al Kassem et al. (2024)	Syrian civil war
Almohamad (2021)	Manbij, Syria
Angiuli et al. (2020)	Nineveh and Nebi Yunus, Iraq
Casana (2015)	Syrian looting
Casana and Laugier (2017)	Syrian civil war
Casana and Panahipour (2014)	Syrian looting
Fradley and Sheldrick (2017)	Gold mining sites, Egypt
Kopij et al. (2024)	Jordan
Lubin and Salem (2019)	Aleppo, Syria
Mamo et al. (2022)	Al-Hasakah, Syria
Marx (2016)	Syrian urban sites
Masini and Lasaponara (2020)	Syrian looting

Fig. 1.2 The Great Colonnade in Apamea, Syria. 2011 (Top) and 2012 (Bottom). 35°25′34.31″N, 36°24′10.44″E. Scale bar = 200 m, ©Google Earth

world, with an estimated seating capacity of more than 20,000 people. The population of Apamea in 6 AD was at least 130,000 but was maybe as high as 500,000. As a result of the Syrian Civil War the ancient city has been damaged and looted by treasure hunters seeking ancient artefacts, and has become covered in a lunar landscape of hundreds of holes. The war-torn population of Qalaat al-Madiq and the surrounding communities, faced with few if any positive economic choices, turned to the looting of antiquities during the social and economic breakdown of the conflict. The image shows the site in 2011, and then again in 2012.

No less distressing was the damage inflicted on the great ancient city of Palmyra (Fig. 1.3), a World Heritage Site since 1980 (https://whc.unesco.org/en/list/23/) (Accessed 4th April 2025). This lies in the centre of modern Syria and is an oasis in the Syrian Desert. The city grew wealthy from trade caravans, for the inhabitants became merchants who established colonies along the Silk Road and operated throughout the Roman Empire. Palmyra's wealth enabled the construction of monumental projects, such as the Great Colonnade, the Temple of Bel (Fig. 1.3), and distinctive tower tombs. By the third century it had become a prosperous regional centre, reaching the apex of its power in the 260s AD.

On 23 May 2015, Islamic State (IS) militants destroyed various statues. They destroyed the Temple of Baalshamin on 23 August, and then on 30 August, IS destroyed the Temple of Bel. IS had also destroyed three of the best-preserved tower tombs, a monumental arch, the tetrapylon, and part of the theatre. As Harrowell (2016) has shown, the representatives of IS have explained their vandalism (and looting) of ancient sites as a response to the idolatry and forbidden images they perceived there, but they also sought to erase the physical evidence that a multicultural society once thrived in Syria, denying a history that runs counter to their beliefs. Figure 1.3

Fig. 1.3 Palmyra, Syria, Temple of Bel, 2010 (Top) and 2016 (Middle). 34°32′51.20″N, 38°16′28.07″E. Scale bar = 100 m, ©Google Earth. Bottom: Temple of Bel, prior to its destruction (https://commons.wikimedia.org/wiki/File:Temple_of_Bel,_Palmyra,_Syria_-_1.jpg). (Accessed 4th April 2025). Public domain

shows the Temple of Bel as it was in 2010, before the Civil War, and in 2016 after IS destruction had been achieved.

Another badly destroyed Syrian site is Mari (Fig. 1.4). This was home to Zimri-Lim's extraordinary palace, and this dates to around 2500–2300 BC. Mari is probably the site that has provided the greatest amount of information useful for understanding the history between the third and the end of the second millennia in Syro-Mesopotamia (https://whc.unesco.org/en/tentativelists/5702) (Accessed 4th April 2025). Its situation between the two seas, the 'rising sun' and the 'setting sun'—the Persian Gulf and the Mediterranean—made Mari an almost-obligatory transit point. Because it was founded around 2900 BC, just at the start of the expansion of urbanisation, Mari allows the study of the changes that occurred in the rapid evolution from a rural way of life to an urban one in Mesopotamia (see Scarre et al. 2021, p. 221).

When IS emerged, Mari was in one of the first areas to fall under its control in early 2014. Under IS the site suffered an immediate explosion of looting; and remote sensing has revealed the change from archaeological site to lunar landscape in a matter of months. More than 1,500 new looting pits were recorded at Mari between 2013 and 2015. These

Fig. 1.4 Mari, Syria. 34°33′6.22″N, 40°53′18.63″E. Scale bar = 100 m, ©Google Earth. Top: 2011, Bottom: 2018

probably represent the removal of a huge quantity of ancient objects, sold into the illegal antiquities market to fund IS's war activities.

Other Syrian sites suffered damage during the Civil War as a result of the conflict between the Syrian government forces and the rebels. Damaged sites included the Citadel in Aleppo (see Chap. 4) (Bandarin 2022) and also the Great (Umayyad) Mosque (Fig. 1.5). This was built in the beginning of the eighth century AD with much of the current structure dating back to the eleventh through to the fourth centuries. The Minaret and much else were destroyed in 2013, as well as surrounding buildings. The Great Mosque was part of a World Heritage Site (https://whc.unesco.org/en/list/21/) (Accessed 4th April 2025) and is now in the process of restoration.

In Iraq, IS also played a role (https://www.unitad.un.org/sites/www.unitad.un.org/files/20240914_dch_public_final_2.pdf) (Accessed 4th April 2025), notably in the destruction of the Ziggurat of Nimrud (Fig. 1.6). Nimrud was the capital of the ancient Assyrian civilization and the 43 m tall temple was at the centre of its spiritual life. Built about 2,900 years ago by King Ashurnasirpal II (883–859 BC), the mud brick structure was dedicated to Ninurta, a god of war and the city's patron deity. In 2015, IS announced its intention to destroy the site because of its 'un-Islamic' Assyrian nature. It employed bulldozers to do so. Their tracks are clearly evident in the GE image.

Creative commons Also in Iraq is the site of Tel Afar in Nineveh Governorate (Fig. 1.7). Its citadel was built by the Ottomans, although it contains remains dating back to the Assyrian period. Tel Afar fell to IS in June 2014, and the militants used the citadel as a prison for women and girls who were to be forcibly married to IS members. In December of that year IS blew up the city's citadel and northern and western walls, causing extensive damage

Fig. 1.5 The Great Mosque in Aleppo, Syria. 36°11′58.27″N, 37° 9′25.22″E. Scale bar = 60 m, ©Google Earth. Top: in 2012. Bottom: in 2024

(https://kirkuknow.com/en/news/68395) (Accessed 4th April 2025). The militants also excavated some of the ruins within the citadel, probably to look for antiquities which they could sell. The overall extent of damage to the Citadel was 91–100 percent (https://www.unitad.un.org/sites/www.unitad.un.org/files/20240914_dch_public_final_2.pdf) (Accessed 4th April 2025).

In addition, many archaeological sites in Gaza have been obliterated by rampant Israeli forces (Taha 2024). UNESCO is currently attempting to assess the appalling damage (sometimes now being called 'urbicide') using satellite images (https://www.unesco.org/en/gaza/assessment) (Accessed 4th April 2025), and has reported that 'as of 17 September 2024, out of the 120 sites that UNESCO has been able to assess through satellite imagery, 69 sites have been damaged since 7 October 2023—10 religious sites, 43 buildings of historical and/or artistic interest, 2 depositories of movable cultural property, 6 monuments, 1 museum, and 7 archaeological sites' (Fig. 1.8). Amongst the archaeological sites destroyed by bombing and bulldozing was the ancient port of Anthedon, a World Heritage Tentative Site (https://www.worldheritagesite.org/tentative/id/5719) (Accessed 4th April 2025).

Putin's dreadful war in Ukraine has also caused the destruction of some sites, such as the Sviatohirsk Cave Monastery in the Donbas (49° 1′42.66″N, 37°34′2.36″E). In 2022 GE shows its majesty and superb state of repair, but now it is in ruins, primarily as a result of Russian shelling, air attacks, and fire (Fig. 1.9).

It is worth remembering the human tragedies that all these conflicts have caused. As Meskell (2018, p. 183) remarked:

Fig. 1.6 Ziggurat of Nimrud, Iraq. Top: Before destruction. Middle: Showing it after its destruction by bulldozers in 2016. 36° 6′1.73″N, 43°19′45.13″E. Scale bar = 100 m, ©Google Earth. Bottom: The cruel King Ashurnasirpal II, with fine hair, earring and other ornamentation. https://en.wikipedia.org/wiki/Ashurnasirpal_II#/media/File:Middle_East-_G9_Assyria_Nineveh_-_44841858665.jpg. (Accessed 4th April 2025)

> The danger here is intervening in the rescue of a purely material heritage, often a Classical antiquity for which Western nations conside themselves to be the legitimate inheritors, while not fully addressing those living and dying through the ongoing conflict or he resultant refugee crisis…

War and bigotry have not been the only causes of site destruction in the Middle East and elsewhere. Also important has been the spread of buildings and roads across them. This can be illustrated by images from Jabal Be'ih, Raqqa, Syria, where in 2013 there were no modern buildings, while in the next decade they had spread in large numbers across the site (Fig. 1.10).

In Murzuk, Libya, the spread of modern buildings has almost obliterated the old town (Fig. 1.11). It has expanded substantially, particularly since 2011. Large areas on the outskirts of the town have been bulldozed, levelling post-medieval buildings, cemeteries and field systems to provide room for new housing. Between 2012 and 2016, large parts of the post-medieval town, which was abandoned at some point before the nineteenth century, have been bulldozed,

Fig. 1.7 Tel Afar, Iraq, in 2014 (top) and 2016 (bottom). 36°22′27.93″N, 42°27′16.51″E. Scale bar = 100 m, ©Google Earth

and the fifteenth- or sixteenth-century wall circuit with its D-shaped towers has almost entirely disappeared (Rayne et al. 2017a). There has also been encroachment on Cyrene in Libya (Abdulkariem and Bennett 2014), as seen on GE (Menozzi et al. 2019). GE has also been used to assess the destruction of archaeological sites in Egypt as a result of urbanisation (Abdelfattah 2025).

Another location where severe site destruction has occurred is Afghanistan, following the takeover of power by the Taliban in 2021 (https://www.science.org/content/article/looters-continue-pillage-afghanistan-s-rich-archaeological-heritage) (Accessed 4th April 2025), though earlier pillage was undertaken at sites like Dilberjin Tepe by IS. Delbarjin is the remains of an ancient city located 40 km northwest of Balkh, possibly built during the Achaemenid Empire in the fifth century BC. By comparing images from 2011 and 2022, it is evident that widespread looting and bulldozing has taken place and that a road has been inserted into the site from the east (Fig. 1.12).

Similarly severe changes have been wrought at Zadian (Fig. 1.13), also in northern Afghanistan, where bulldozers have been highly active (https://eo4heritage.esa.int/iframe-agenda/files/ID124_Fafet.pdf) (Accessed 4th April 2025) and at Dashli, where a series of buildings dating back to the Chalcolithic (https://pages.vassar.edu/central-asia-sites/2017/05/10/dashly/) (Accessed 4th April 2025) appears to have been demolished by bulldozing since 2021 as shown on GE (https://culturalpropertynews.org/afghanistan-sites-demolished-by-bulldozers/https://culturalpropertynews.org/afghanistan-sites-demolished-by-bulldozers/)

Fig. 1.8 Northern Gaza Strip. 31°32′39.52″N, 34°27′17.81″E. Scale bar = 300 m, ©Google Earth. Top: 8th October 2023. Bottom: 24th November 2023

(Accessed 4th April 2025). Such bulldozing may be done to produce building aggregates or to expose potential valuable artefacts.

GE has also been successfully used to monitor looting in Egypt (Parcak et al. 2016), Peru (Contreras 2010), Cyprus (Agapiou et al. 2017), northwest China (Caspari 2018), Jordan (Kopij et al. 2024), and Syria (Lasaponara and Masini 2018). In Syria, the classical site of Dura Europus shows the presence of huge numbers of looting pits (Fig. 1.14) made by IS and earlier diggers (Casana and Laugier 2017).

Other threats to heritage sites include coastal erosion and accelerating sea-level rise associated with global heating (Reimann et al. 2018; Vousdoukas et al. 2022; Erlandson et al. 2025; Pollard et al. 2025). GE can be used to identify vulnerable sites. One of these is Sabratha in Libya (Nikolaus et al. 2023) (Fig. 1.15).

GE has also been used to study sand encroachment around the pyramids at Meroe in Sudan (Fahmy et al. 2023) and potential and actual damage to sites along the new TAPI gas pipeline in Afghanistan (Rouhani et al. 2024). In South America, construction of the Pan-American Highway and off-road driving have caused great damage in proximity to the Nazca (Nasca) Lines geoglyphs (Fig. 1.16), and this can be clearly seen on GE and other remote sensing platforms (Tapete et al. 2013; Chapman et al. 2015). Stone (desert) pavements are especially prone to damage by such

Fig. 1.9 Top: Sviatohirsk Cave Monastery, Ukraine, in 2022, prior to shelling. 49° 1′42.66″N, 37°34′2.36″E. Scale bar = 70 m, ©Google Earth. Bottom: Current State (https://commons.wikimedia.org/wiki/File:St._George%27s_Skete_of_Sviatohirsk_Lavra_after_Russian_shelling_(01).jpg) (Accessed 4th April 2025)

activities, and due to their specific characteristics, geoglyphs are very fragile because they can be severely damaged or completely destroyed by simply stepping on them (Masini and Lasaponara 2015).

Earthquakes can also cause damage to heritage sites. This was made evident in Hatay, southeast Türkiye, after the 6 February 2023 earthquake. This caused significant damage, inter alia, to the great sixth-century Ulu Mosque in Adıyaman. After the earthquake, only the walls on the north and west sides were standing together with the part of the roof on the northwest of the mosque (Boyoğlu et al. 2023) (Fig. 1.17).

Fig. 1.10 Jabal Be'ih, Raqqa, Syria. Before (Top, 2013) and After (Bottom, 2023) Urbanisation. 35°57′26.20″N, 39° 2′49.55″E. Scale bar = 100 m, ©Google Earth

Fig. 1.11 Top: The old town of Murzuk, Libya, in 2003. Bottom: In 2022. 25°54′33.97″N, 13°54′45.62″E. Scale bar = 100 m, ©Google Earth

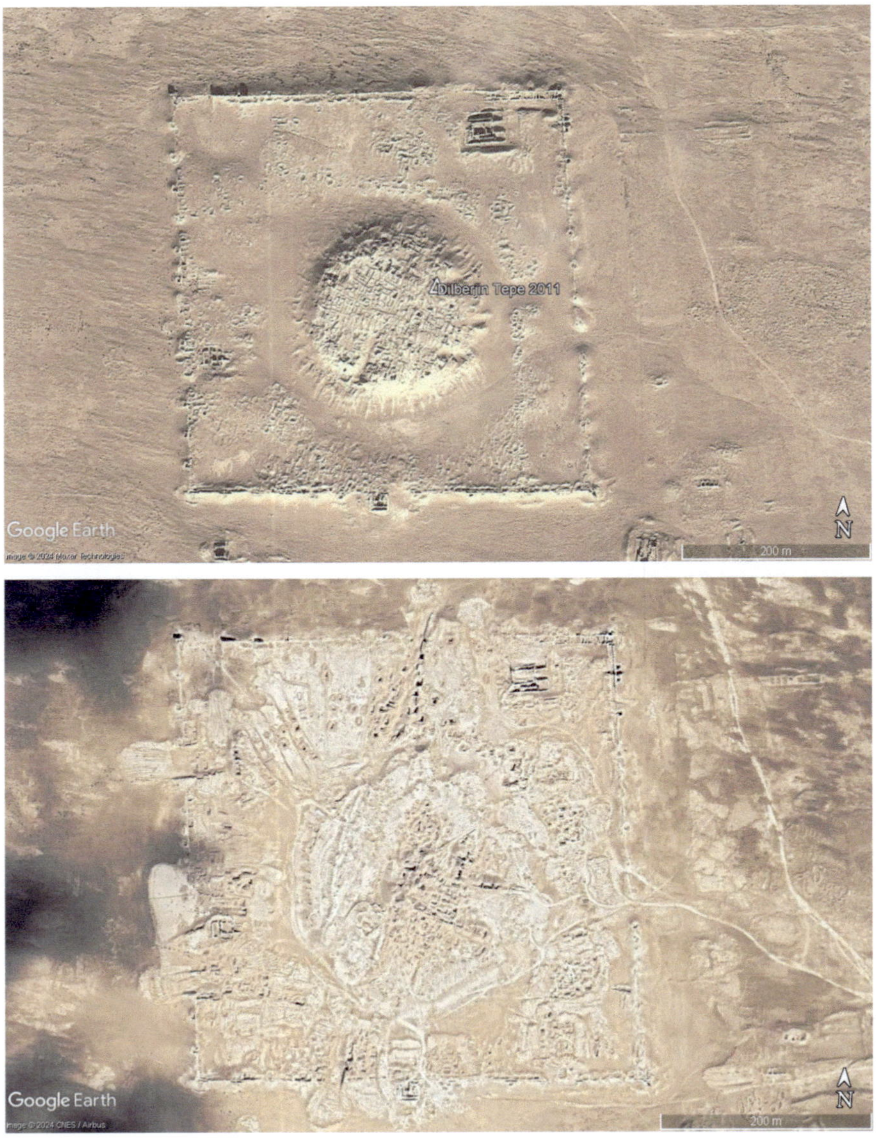

Fig. 1.12 Top: Diljerbin Tepe, Afghanistan in 2011. 37° 1′22.13″N, 66°31′36.10″E. Scale bar = 200 m, ©Google Earth. Bottom: In 2022

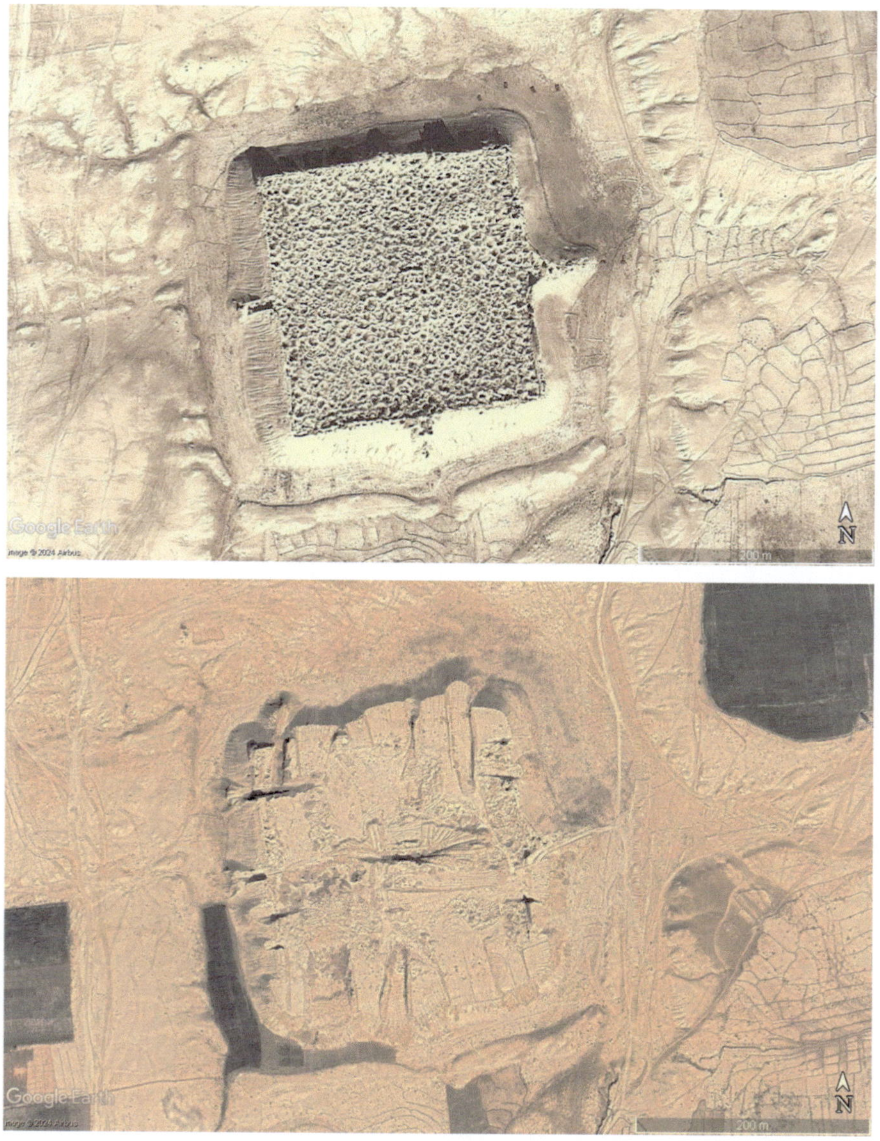

Fig. 1.13 Zadian, Afghanistan. 37°2′46.20″N, 66°56′42.29″E, Scale bar = 200 m, ©Google Earth. Top: in 2013, Bottom: In 2023

Fig. 1.14 Dura Europus, Syria, showing huge numbers of looting pits. 34°44′50.19″N, 40°43′28.21″E. Scale bar = 100 m, ©Google Earth

Fig. 1.15 Sabratha, Libya. Coastal erosion of an archaeological site. 32°48′25.98″N, 12°28′55.09″E. Scale bar = 70 m, ©Google Earth

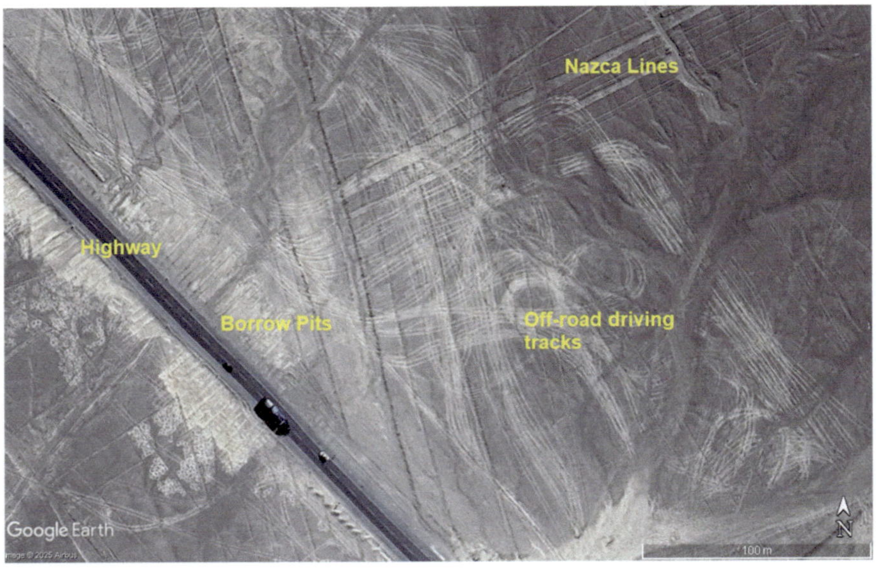

Fig. 1.16 The Nazca lines, Peru, showing damage caused by road construction and by off-road driving. 14°41′27.21″S, 75° 6′55.82″W. Scale bar = 100 m, ©Google Earth

Fig. 1.17 Ulu Mosque, Hatay, Türkiye. 36°12′7.25″N, 36° 9′42.96″E. Scale bar = 80 m, ©Google Earth. Top: in 2022, prior to 2023 earthquake. Bottom: in 2024, after the 2023 earthquake

References

Abdelfattah KA (2025) The vanishing heritage: An in-depth study of disappearing archaeological sites in Kafr El-Sheikh, Egypt. J Assoc Arab Univ Tour Hosp 28(1):141–170

Abdulkariem A, Bennett P (2014) Libyan heritage under threat: the case of Cyrene. Libyan Stud 45:155–161

Agapiou A, Lysandrou V, Hadjimitsis DG (2017) Optical remote sensing potentials for looting detection. Geosciences 7(4):98

Al Kassem A (2024) Assessment of the impact of the Syrian conflict on archaeological sites in the Daraa region. Levant 56(3):425–446

Almohamad A (2021) The destruction and looting of cultural heritage sites by ISIS in Syria: The case of Manbij and its countryside. Int J Cult Prop 28(2):221–260

Angiuli E, Pecharromán E, Vega Ezquieta P, Gorzynska M, Ovejanu I (2020) Satellite imagery-based damage assessment on Nineveh and Nebi Yunus archaeological site in Iraq. Remote Sens 12(10):1672

Badman T (2010) World heritage and geomorphology. In: Migoń P (ed) Geomorphological landscapes of the world. Springer, Dordrecht, pp 357–368

Bandarin F (2022) The destruction of Aleppo: the impact of the Syrian war on a Palmyra city. In: Cuno J, Weiss TG (eds) Cultural heritage and mass atrocities. Getty Publications, Los Angeles, pp 186–201

Boyoğlu CS, Chike I, Caspari G, Balz T (2023) Assessing the impact of the 2023 Kahramanmaras earthquake on cultural heritage sites using high-resolution SAR images. Heritage 6(10):6669–6690

Casana J (2015) Satellite imagery-based analysis of archaeological looting in Syria. Near East Archaeol 78(3):142–152

Casana J, Panahipour M (2014) Satellite-based monitoring of looting and damage to archaeological sites in Syria. J East Mediterr Archaeol Herit Stud 2(2):128–151

Casana J, Laugier EJ (2017) Satellite imagery-based monitoring of archaeological site damage in the Syrian civil war. PloS one 12(11):e0188589

Caspari G (2018) Assessing looting from space: the destruction of early iron age burials in northern Xinjiang. Heritage 1(2):21

Chapman BD, Comer DC, Isla JA, Silverman H (2015) The measurement by airborne synthetic aperture radar (SAR) of disturbance within the Nasca world heritage site. Conserv Manag Archaeol Sites 17(3):270–286

Contreras DA (2010) Huaqueros and remote sensing imagery: assessing looting damage in the Virú Valley, Peru. Antiquity 84:544–555

Crawford OGS, Keiller A (1928) Wessex from the air. Oxford Univ Press, Oxford, p 385

Darvill T (1996) Prehistoric Britain from the air. Cambridge Univ Press, Cambridge, p 283

Demirci A, Karaburun A, Kılar H (2013) Using Google Earth as an educational tool in secondary school geography lessons. Int Res Geogr Environ Ed 22(4):277–290

Dolliver HA (2012) Using google earth to teach geomorphology. Geol Soc Am Spec Pap 492:419–429
Eidshaug JS, Bjerck HB, Breivik HM, Risbøl O, Tivoli AM, Zangrando AF (2024) From earth to sky: large-scale archaeological settlement patterns in southernmost South America based on ground surveys, UAV LiDAR, and open access satellite imagery. J Isl Coast Archaeol. https://doi.org/10.1080/15564894.2024.2426828
Erlandson JM et al (2025) Rising seas endanger maritime heritage. Science 387:482
Fahmy A, Domínguez-Bella S, Martínez-López J, Molina-Piernas E (2023) Sand dune movement and flooding risk analysis for the pyramids of Meroe, Al Bagrawiya archaeological site, Sudan. Herit Sci 11(1):136
Fisher GB, Amos CB, Bookhagen B, Burbank DW, Godard V, Whitmeyer SJ (2012) Channel widths, landslides, faults, and beyond: the new world order of high-spatial resolution google earth imagery in the study of earth surface processes. Geol Soc Am Spec Pap 492:1–22
Fradley M, Sheldrick N (2017) Satellite imagery and heritage damage in Egypt: a response to Parcak et al. (2016). Antiquity 91:784–792
Frere SS, St Joseph JKS (1983) Roman Britain from the air. Cambridge Univ Press, Cambridge, p 232
Goudie AS (2022) Desert landscapes of the world with google earth. Springer, Cham, p 271
Goudie AS (2023) Landscapes of the anthropocene with google earth. Springer, Cham, p 249
Goudie AS (2025) Rocky landscapes of the world with google earth. Springer, Cham, p 262
Harkel A, Bewley R (2018) An incredible journey? Understanding ancient landscapes from England to the Middle East and North Africa. Mediev Settl Res 33:40–59
Harrowell E (2016) Looking for the future in the rubble of Palmyra: destruction, reconstruction and identity. Geoforum 69:81–83
Hauser K (2008) Bloody old Britain. Granta Books, London, p 286
Kaimaris D, Georgoula O, Patias P, Stylianidis E (2011) Comparative analysis on the archaeological content of imagery from google earth. J Cult Herit 12(3):263–269
Kennedy D, Bishop MC (2011) Google earth and the archaeology of Saudi Arabia. A case study from the Jeddah area. J Archaeol Sci 38(6):1284–1293
Kennedy D, Bewley R (2004) Ancient Jordan from the air. British Academy, London, p 282
Kennedy D, Riley D (1990) Rome's desert frontier from the air. Batsford, London, p 256
Kopij K, Ćwiąkała P, Puniach E, Sochacki G, Miszk Ł, Bodzek J (2024) Temporal analysis of looting activity in Tūwāneh (Southern Jordan). J Field Archaeol 49(1):74–94
Lasaponara R, Masini N (2018) Space-based identification of archaeological illegal excavations and a new automatic method for looting feature extraction in desert areas. Surv Geophys 39:1323–1346
Liang J, Gong J, Li, W (2018) Applications and impacts of google earth: a decadal review (2006–2016). ISPRS J Photogram Remote Sens 146:91–107
Lindsay I, Mkrtchyan A (2023) Free and low-cost aerial remote sensing in archaeology: an overview of data sources and recent applications in the South Caucasus. Adv Archaeol Pract 11(2):164–183
Lisle RJ (2006) Google earth: a new geological resource. Geol Today 22(1):29–32
Lubin A, Saleem A (2019) Remote sensing-based mapping of the destruction to Aleppo during the Syrian civil war between 2011 and 2017. Appl Geogr 108:30–38
Luo L et al (2018) Google earth as a powerful tool for archaeological and cultural heritage applications: a review. Remote Sens 10(10):1558
Mamo AR, Ibraheem IM, Al Kassem A, Al-Khalil A, Hopper K (2022) The impact of the Syrian conflict on archaeological sites in Al-Hasakah province. J Archaeol Sci Rep 43:103486
Marx AJ (2016) Detecting urban destruction in Syria: a Landsat-based approach. Remote Sens Appl Soc Environ 4:30–36
Masini N, Lasaponara R (2015) Satellite and close range analysis for the surveillance and knowledge improvement of the Nasca geoglyphs. Remote Sens Environ 236:111447
Masini N, Lasaponara R (2020) Recent and past archaeological looting by satellite remote sensing: approach and application in Syria. In: Hadjimitsis DG et al (eds) Remote sensing for archaeology and cultural landscapes: best practices and perspectives across Europe and the Middle East. Springer, Cham, pp 123–137
Menozzi O, Cherstich L, Di Valerio E, Di Antonio MG, Saad AA (2019) A proposal for the use of core and buffer zones to protect the heritage of the Cyrenean landscapes. Libyan Stud 50:107–135
Meskell L (2018) A future in ruins. UNESCO, world heritage and the dream of peace. Oxford Univ Press, Oxford, p 372
Nikolaus J, Westley K, Breen C (2023) Endangered maritime archaeology in North Africa—the MarEA project. Libyan Stud 54:135–150
O'Reilly DJ, Scott G (2015) Moated sites of the iron age in the Mun River Valley, Thailand: new discoveries using google earth. Archaeol Res Asia 3:9–18
Ortiz-Guerrero C, Loizzo J (2024) Rocks really rock: electronic field trips via web google earth can generate positive impacts in attitudes toward earth sciences in middle-and high-school students. Geosci Commun 7(2):101–119
Parcak SH (2009) Satellite remote sensing for archaeology. Routledge, London p, p 308
Parcak S, Gathings D, Childs C, Mumford G, Cline E (2016) Satellite evidence of archaeological site looting in Egypt: 2002–2013. Antiquity 90(349):188–205
Pollard E et al (2025) Climate change, coastal heritage digitization, and local community engagement at the ruins of Kilwa Kisiwani World Heritage Site, Tanzania. J Field Archaeol 50(1):6–21
Rayne L, Sheldrick N, Nikolaus J (2017a) Endangered archaeology in Libya: recording damage and destruction. Libyan Stud 48:23–49
Rayne L, Bradbury J, Mattingly D, Philip G, Bewley R, Wilson A (2017b) From above and on the ground: geospatial methods for recording endangered archaeology in the Middle East and North Africa. Geosciences 7(4):100
Reimann L, Vafeidis AT, Brown S, Hinkel J, Tol RS (2018) Mediterranean UNESCO world heritage at risk from coastal flooding and erosion due to sea-level rise. Nature Comm 9(1):4161
Riley DN (1987) Air photography and archaeology. Duckworth, London, p 152
Rouhani B, Tarabay B, Jabri L, Saroufim M (2024) Report on the cultural heritage survey for the TAPI Project, Afghanistan: Pilot phase, EAMENA-ICONEM July 2024. EAMENA, School of Archaeology, University of Oxford. Accessed 4 Apr 2025. https://doi.org/10.5281/zenodo.13912379
Royal Commission on Historical Monuments (England) (1970) An inventory of the historical monuments in the County of Dorset, vol 3, part 2, p 385. HMSO, London and Woking
Sadr K, Rodier X (2012) Google Earth, GIS and stone-walled structures in southern Gauteng, South Africa. J Archaeol Sci 39(4):1034–1042
Scarre C (ed) (2005) The human past. Thames and Hudson, London, p 784
Scarre C, Fagan BM, Golden C (2021) Ancient civilizations, 5th edn. Routledge, London, p 686
Scheffers AM, May SM, Kelletat DH (2015) Landforms of the world with google earth: understanding our environment. Springer, Dordrecht, p 401
Taha H (2024) Destruction of cultural heritage in Gaza. Jerusalem Quarterly 97:45–70
Tapete D, Cigna F, Masini N, Lasaponara R (2013) Prospection and monitoring of the archaeological heritage of Nasca, Peru, with ENVISAT ASAR. Archaeol Prospect 20(2):133–147

Thomas DC, Kidd FJ, Nikolovski S, Zipfel C (2008) The archaeological sites of Afghanistan in google earth. AARG News 37:22–30

Tooth S (2015) Google earth as a resource. Geography 100:51–56

Traganos D, Aggarwal B, Poursanidis D, Topouzelis K, Chrysoulakis N, Reinartz P (2018) Towards global-scale seagrass mapping and monitoring using Sentinel-2 on Google Earth Engine: the case study of the Aegean and Ionian Seas. Remote Sens 10(8):1227

Vousdoukas MI et al (2022) African heritage sites threatened as sea-level rise accelerates. Nat Clim Chang 12(3):256–262

Williams CM, Baker GS, Ault BA (2012) Enhancing usability of near-surface geophysical data in archaeological surveys via google earth. Geol Soc Am Spec Pap 492:49–62

Landscapes of Water Management

2

Abstract

Water management of surface and groundwater resources has a long history and has had many landscape impacts. Phenomena include terraces across stream channels, canals, dams, reservoirs, irrigation systems, *qanat*, *puquios,* aqueducts and stone fish weirs. Some of these features date back to the Neolithic.

Keywords

Terraces · Canals · Irrigation · Groundwater · Dams · Aqueducts · Fish weirs

2.1 Introduction

The interventions of humans over millennia have produced archaeological and built heritage sites and features which are large enough and persistent enough to be clearly visible in the landscape from GE. For thousands of years, many landscapes have developed as a result of various types of water management schemes. These include terraces across stream channels, canals, dams, reservoirs, irrigation systems, *qanat*, *puquios,* aqueducts, and stone fish weirs. A good summary and exemplification of some water management techniques in England is provided by Cook and Williamson (1999).

2.2 Terraced Stream Courses

Some terrace structures have been built for harvesting and control of flood waters (Beckers et al. 2013). Terraced wadis (desert drainage lines) exist in Jordan and there is evidence of ancient floodwater farming (McCorriston and Oches 2001). Stone terrace walls were also built in valleys in the Negev Desert (Fig. 2.1). Many thousands of them exist (Ore and Bruins 2012). Such terracing dates back to the Neolithic and Chalcolithic (a transitional phase between the Neolithic and Bronze Ages) but it was also practised from the Bronze Age through to Byzantine times (Bruins 2012). Ancient walls, dams and terraces for flood control, flood water irrigation and silt trapping also occur widely in the Baluchistan, Kohistan and Sindh regions of Pakistan, where they are called *gabar-bands* (Fig. 2.2) (Raikes 1964; Harvey and Flam 1993) or *sailaba* (Kahlown and Hamilton 1996; Brunner 2006). In the arid southwest of America, pre-Columbian Indians also made extensive use of floodwater farming (Bryan 1929).

Somewhat similar features occur in South India, where they are called *tanks.* Here, the landscape consists of myriads of little streams and areas of overland flow that have been dammed by small crescent-shaped earth dams called *bunds* to give small lakes which typically have an area of c 5 ha (Fig. 2.3). In India as a whole the number of tanks is reported to be between 0.2 and 0.35 million, though the number in use has declined in recent decades as farmers have turned to groundwater pumping to produce irrigation water for their crops (Reddy et al. 2018). Tanks also occur in profusion in parts of Sri Lanka, where they are called *wewa.*

2.3 Canals

Canals are humanly made rivers. They are artificial waterways or engineered channels built for drainage management (e.g. flood control and irrigation), for conveying ships, boats, and barges, and occasionally, to provide fisheries. A canal can be created where no stream presently exists. Either the body of the canal is excavated or the sides of the canal are created by making dykes or levees by piling stone, concrete, or other materials. The water for the canal has to be provided from an external source, such as streams or reservoirs.

A. Goudie, *Landscape Heritage of the World with Google Earth*, https://doi.org/10.1007/978-3-031-95555-6_2

Fig. 2.1 Avdat Wadi Terracing, Israel. 30°48′18.12″N, 34°46′3.81″E. Scale bar = 200 m, ©Google Earth

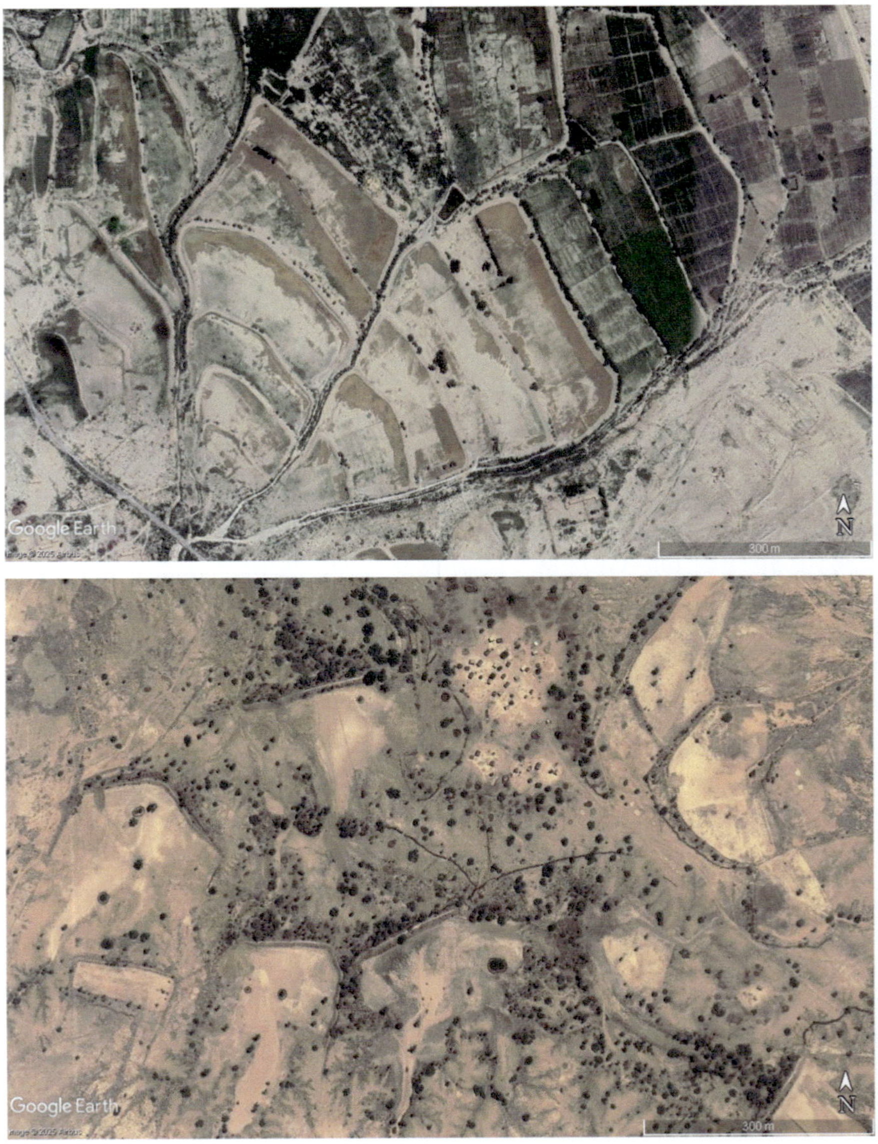

Fig. 2.2 Gabarbands (sailaba) in Baluchistan, Pakistan. Top: Wadh, 27°20′6.53″N, 66°19′4.54″E. Scale bar = 300 m, ©Google Earth. Bottom: Phang, 25°48′49.60″N, 67°20′38.79″E. Scale bar = 300 m, ©Google Earth

Fig. 2.3 Tank landscape of Southeast India. 9°44′20.71″N, 78°54′37.44″E. Scale bar = 7 km, ©Google Earth

The oldest known canals were built for irrigation in Mesopotamia c 6000 years ago. In India, the Harappan civilization (c 5000 years ago) had sophisticated irrigation and storage systems developed, including reservoirs built at Girnar and at Harappa itself (see Chap. 7). In Egypt, canals date back at least to the time of Pepi I Merye (2332–2283 BC), who ordered a canal to be built to bypass the cataract on the Nile near Aswan.

There were also extensive networks of canals near Phoenix in Arizona, built by Hohokam peoples. The Hohokam, who occupied large areas along the Salt and Middle Gila Rivers (Fig. 2.4), are believed to have emerged from 450 AD onwards and flourished in the period of 700–1150 AD. Around 1250 AD they underwent a significant social reorganisation that eventually collapsed around 1400 to 1450 AD. They used the waters of the Salt and Gila Rivers to build an assortment of simple canals with weirs for agriculture from 800 to 1400 AD (Woodson et al. 2015; Tianduowa et al. 2018; Woodson 2024). Archaeologists have now identified 19 irrigation systems along the Middle Gila River.

In Central America the Mayans built canals for agricultural and fishing purposes, and Harrison-Buck et al. (2024) mapped some of these using GE.

In Cambodia, the economy of the great city of Angkor Wat was based on rice cultivation enabled by a large system of canals and reservoirs (Fletcher et al. 2008). These are clearly visible on GE at 13°29′40.74″N, 103°51′42.13″E.

The Grand Canal of China (Fig. 2.5) is still the longest canal in the world today and the oldest extant one (Tan et al. 2019). It is 1,794 km long. Starting in Beijing, it

Fig. 2.4 Canals, Gila River, Arizona, USA. 33°10′56.35″N, 111°57′36.23″W. Scale bar = 200 m. ©Google Earth

Fig. 2.5 Grand Canal, China. 30°24′9.17″N, 120° 7′38.36″E. Scale bar = 1 km, ©Google Earth

passes through Tianjin and the provinces of Hebei, Shandong, Jiangsu, and Zhejiang to the city of Hangzhou, linking the Yellow and Yangtze rivers. The oldest parts of the canal date back to the fifth century BC (Sui Dynasty), and reached a peak in the thirteenth century (Yuan Dynasty). However, only the section from Hangzhou to Liangshan County is currently navigable. It is a World Heritage Site (https://whc.unesco.org/en/list/1443/) (Accessed 4th April 2025).

Some canals were dug for land drainage. Notable examples in Britain include those in the Somerset Levels (Williams 2009) and the English Fens (Darby 1940) (Fig. 2.6). In the case of the latter some very large channels were dug in the seventeenth century AD to drain a low-lying flat marshy, and flood-prone area, so that it could be developed for farming. Two cuts were made in the Cambridgeshire Fens to join the River Great Ouse to the North Sea at King's Lynn. These were called the Old Bedford River and the New Bedford River. Both of these cuts were named after the fourth Earl of Bedford who, along with some *gentlemen adventurers*, funded the construction and were rewarded with large grants of the resulting farmland. The work was directed by engineers from the Netherlands, including the great Cornelius Vermuyden. The two Bedford Rivers stretched 33.5 km from Earith to Denver Sluice. General discussions of Medieval and post-Medieval drainage of marsh and fen are provided by Silvester (1999) and Taylor (1999), respectively. Rippon (2000) provides a penetrating discussion of the Roman and Medieval transformations of coastal wetlands in the UK and northwest Europe.

Some lakes were also drained to make land available for agriculture. A very early example of this is Lake Copais in central Greece. This was partially drained during the thirteenth-century BC (Mamassis et al. 2015). Other later examples discussed by Goudie (2023, Sect. 7.2) include Lake Fucine in Italy (Fig. 2.7) and Lake Texcoco in Mexico.

2.4 Dams

The construction of dams dates back to prehistoric times, particularly in Mesopotamia and the Middle East (Angelakis et al. 2024). The earliest known dam is the Jawa Dam in Jordan. It was comprised of an originally 9 m high and 1 m wide stone wall, supported by a 50 m wide earthen rampart. The structure is dated to c 5000 years ago. In China one of the world's oldest examples of large-scale and formalised water management, involving dam construction, was that of the Liangzhu culture in the Yangtze Delta, dated at 5,300–4,300 years BP (Liu et al. 2017). The Ancient Egyptians built a dam at Wadi Al-Garawi, about 25 km south of Cairo. This was 102 m long at its base. It was built around 4800 years ago as a diversion dam for flood control. In Iran the oldest known dams were built 2200 years ago. In Yemen, a great dam was built at Ma'rib. The initial date of the first construction lies somewhere between 1750 and 1700 BC. It was composed of packed earth, triangular in cross section, and 580 m in length and 4 m high. Around 500 BC, its height was increased to 7 m. In Classical Greece, during the Mycenaean Era (1900–1100 BC), large dams were built in the Peloponnese. One was erected around 1200 BC across the Lakissa River to protect the town of Tyrins from flooding (Angelakis et al. 2020). In the Sanchi area of central India, the earliest phase of dam construction occurred approximately between the third and first centuries BC in keeping with the main building phases at neighbouring Buddhist sites (Shaw et al. 2007). They were built to provide irrigation, probably

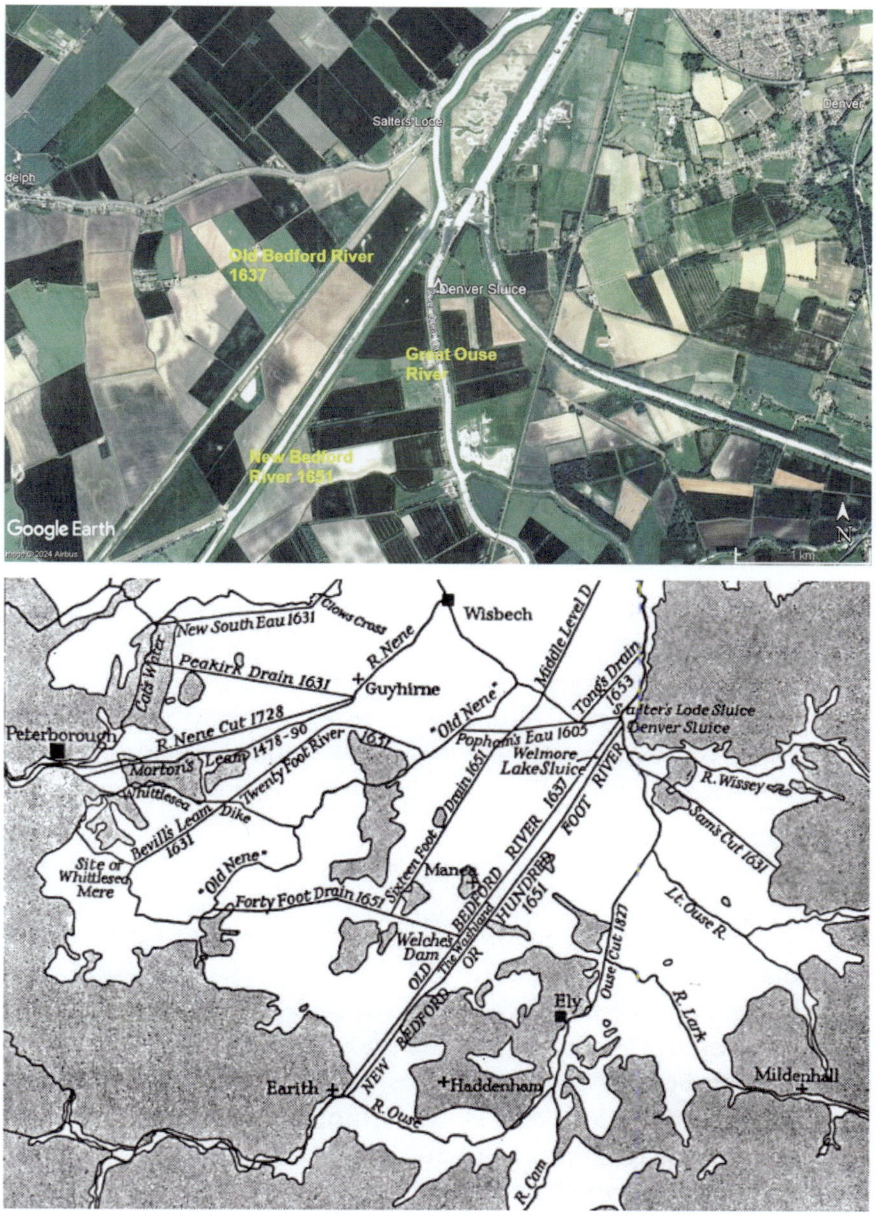

Fig. 2.6 Top: the Fenland of eastern England. 52°34′43.75″N, 0°20′28.06″E. Scale bar = 1 km, ©Google Earth. Bottom: some drainage works in the Fenland as drawn by Darby (1940)

for rice. The Romans were also great dam builders, and many examples are known from Libya (Vita-Finzi 1961).

Harbaqa Dam (Fig. 2.8) lies some 80 km southwest of Palmyra in Syria. It was first constructed in the first century AD. It is impressively large: 345 m long, 20.5 m high, and 18 m thick at the base (Kennedy and Riley 1990, pp 70–71).

The Al Bint (Khaybar) dam is one of the largest ancient dams in Saudi Arabia (Fig. 2.9). It is believed to be of either pre-Islamic origin or from the early Islamic Period. It is constructed out of black basalt rocks, which are abundant in the region. The dam is 30 m high, 215 m wide, and 10 m wide at the base. Although partially collapsed, the standing part of the dam is still 135 m long (https://whc.unesco.org/en/tentativelists/6637/) (Accessed 4th April 2025).

2.5 Groundwater Exploitation

The use of groundwater for irrigation has sometimes been achieved by the traditional technique of *Qanat*, as in large tracts of the Middle East) (Wilkinson 1977), North Africa, and China (Barbaix et al. 2022) (Fig. 2.10). They have a number of other local names, including *foggara* (North Africa), *karez* (Afghanistan), and *khettara* (Morocco). *Qanat* is Arabic for 'conduit', and is the most widely-used term for this type of irrigation system amongst English-speakers. They are underground, gently sloping tunnel wells dug backwards into dryland alluvial fans until the water table is pierced. The water is then transmitted down the tunnels by gravity to be used for irrigation. Cities like Marrakech in Morocco very

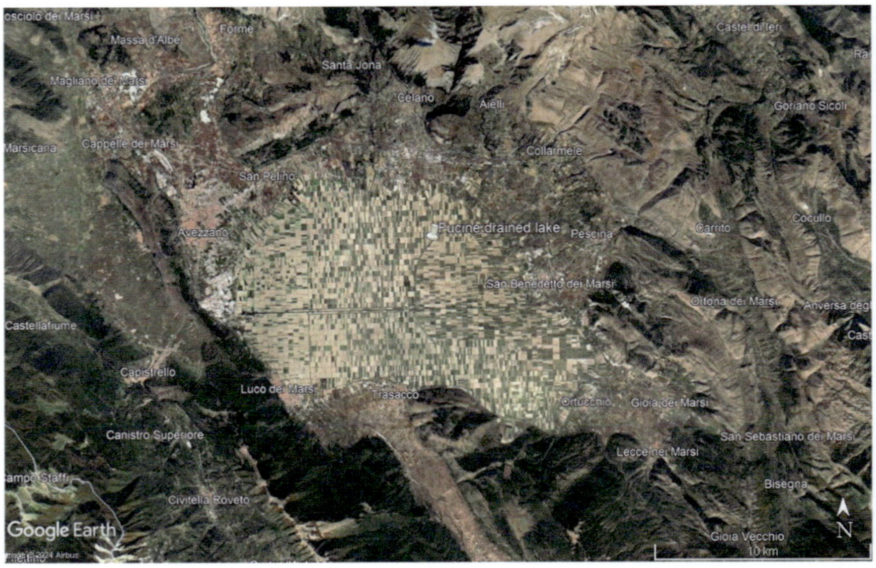

Fig. 2.7 Drained lake Fucine, Italy. 42° 0′19.51″N, 13°33′36.65″E. Scale bar = 10 km, ©Google Earth. Some of the drainage took place in Roman times, but most in the nineteenth century

Fig. 2.8 Harbaqa Dam, Syria. 34°14′42.68″N, 37°37′30.58″E. Scale bar = 300 m, ©Google Earth

Fig. 2.9 Al Bint Dam, Saudi Arabia. 25°29′6.59″N, 39°21′46.21″E. Scale bar = 70 m, ©Google Earth

Fig. 2.10 Top: Qanat, Erfoud, Morocco. 31°25′57.88″N, 4°21′5.56″W. Scale bar = 200 m, ©Google Earth. Bottom: Qanat, Turpan, China. 42°59′3.50″N, 89°38′11.32″E. Scale bar = 100 m, ©Google Earth

much depend on them (Faiz and Ruf 2010). They are characterised by vertical shafts which are constructed for disposal of spoil and to provide air to the miners who made them. These shafts are surrounded by accumulations of excavated debris. In Iran some shafts are 100–120 m deep and some of the tunnels run for as far as 40–50 km. It has often been maintained that these remarkable feats of hydraulic technology originated 2700–2500 years ago in Iran, northern Iraq and eastern Türkiye, and subsequently spread to Arabia (Lightfoot, 2000), Afghanistan, Armenia, Central Asia, northern India, Baluchistan, North Africa, Spain, Mexico, Chile and Peru. However, some have argued for an Arabian origin (see Avni 2018). In some parts of the world they are still in active use, but elsewhere, partly because of the lowering of the water table by modern systems of pumping, they are in decline, as in Morocco, Syria (Lightfoot 1996) and Algeria (Remini et al. 2011). In Iran they have a World Heritage Site designation (https://whc.unesco.org/en/list/1506/) (Accessed 8th June 2023), and there are still over 30,000 qanats operating in that country, but many have been lost as a result of various anthropogenic activities (Maghrebi et al. 2022). GE has proved useful for mapping these features (Luo et al. 2014), not least in Morocco (Hayes-Rich et al. 2023) and in Turpan, China (Li et al. 2022).

Irrigation systems, with dense networks of channels, have been created in arid regions, and some of these are fed by qanat or are led away from springs or major rivers. Figure 2.11 shows an example of irrigated fields in the hyper-arid Kharga Oasis of Egypt. These date back to Ptolemaic times, when groundwater was more plentiful than today (Bravard et al. 2016). The Ptolemaic was the longest and last dynasty of ancient Egypt, lasting from 305 BC until its incorporation into the Roman Republic in 30 BC. Kharga Depression springs provided permanent but decreasing flow during the Holocene

Fig. 2.11 Irrigated fields and canals of Ptolemaic age, El-Deir, Kharga Oasis, Egypt. 25°35′51.71″N, 30°43′38.28″E. Scale bar = 60 m, ©Google Earth

because the deep aquifers have not been refilled by monsoonal rains since the end of the humid Holocene. Locally, wells, known as *Bir*, were dug to reactivate declining spring mound flow when artesian pressure decreased or to tap directly into the aquifers. Some natural springs were still active in the early twentieth century but intensive irrigation and deeper pumping after the onset of British rule in the area (1882) led to the gradual depletion of most remaining natural springs.

In hyper-arid Peru and Chile there are features called *Puquios* (Fig. 2.12). Puquios (from Quecha *pukyu* meaning source, spring, or water well) are ancient systems of subterranean aqueducts. The origin and dating of the Nazca puquios is disputed, although some archaeologists have estimated that their construction began about 500 CE by indigenous people of the Nazca culture. The technology of the puquios is similar to that of the qanats of the old world. The best-known of the puquios are those from Cantalloc in Peru. These are located about 4 km north of the city of Nazca. At intervals of about 10 to 30 m they have ventilation shafts in a typical spiral shape called eyes, through which workpeople could be lowered to carry out cleaning and maintenance work on the system (Lasaponara et al. 2016).

In parts of India, groundwater has been exploited by means of features called stepwells. These were constructed from the seventh to the nineteenth century. They consist of a deep well with stepped sides that permit access to the groundwater at the base, which is an important adaptation to seasonal drought. An important example from Rajasthan is Chand Baori (Fig. 2.13), situated in the village of Abhaneri. Its construction is dated to the eighth-ninth century AD. It is a four-sided well with a large temple located at its back. It has, it is said, 3500 steps that cascade 13 stories deep into a pond at the base and has been constructed in what appears to be an upside-down pyramid-style. Rani Ki Vav stepwell in neighbouring Gujarat is a World

Fig. 2.12 Cantalloc Puquios, Peru. 14°49′35.71″S, 74°54′36.66″W. Scale bar = 70 m, ©Google Earth

Fig. 2.13 Top: Chand Baori stepwell, Rajasthan, India. 27° 0′27.00″N, 76°36′23.61″E. Scale bar = 50 m, ©Google Earth. Bottom: Ground image (https://commons.wikimedia.org/wiki/File:Chand_Baori_perspective_panorama_(July_2022).jpg) (Accessed 4th April 2025)

Heritage Site (https://whc.unesco.org/en/list/922/) (Accessed 4th April 2025).

2.6 Aqueducts

Not all aqueducts are subterranean. Indeed, in many parts of the world, large over-ground systems were created, particularly to bring water to urban centres. The Romans were especially prolific aqueduct constructors. Their aqueducts supplied public baths, latrines, fountains, and private households; they also supported mining operations, milling, farms, and gardens. Rome's first was built in 312 BC, and supplied a water fountain at the city's cattle market. By the third century AD, the city had eleven aqueducts, sustaining a population that may have been over a million people. Their combined conduit length is estimated to have been around 800 km, of which approximately 47 km were carried above ground level on masonry supports. Hundreds of aqueducts were built throughout the Roman Empire. For example, the Zaghouan Aqueduct, over 90 km in length, was built in the second century AD to supply Carthage. Roman Dorchester (Durnovaria) in southern England also had an aqueduct, over 19 km long, bringing water from the River Frome (Putnam 1997; Manley et al. 2024) to what was a substantial city (Wacher 1974, pp. 315–326) with an amphitheatre (Fig. 9.3) called Maumbury Rings.

A particularly well-preserved and striking aqueduct from Roman times is that in Segovia, Spain (Fig. 2.14). It, and the historic old city, have had World Heritage status since 1985 (https://whc.unesco.org/en/list/311/) (Accessed 4th April 2025). It was probably built c 50 AD to bring water from the Río Frío in the Sierra de Guadarrama. It is an enormous construction of masonry, 813 m in length, consisting of four straight segments and two superimposed arcades borne by 128 pillars. At the lowest point of the valley, the aqueduct stands at a height of 28.5 m above ground.

Another strikingly well-preserved Roman aqueduct is the Pont du Gard in southern France (Fonder and Xanthoulis 2007; Ortloff 2020) (Fig. 2.15). It is the major element of a 50 km aqueduct built in the middle of the first century to supply the city of Nîmes, the ancient Roman colony of

Fig. 2.14 Roman Aqueduct, Segovia, Spain. 40°56′49.95″N, 4° 6′57.24″W. Scale bar = 100 m, ©Google Earth

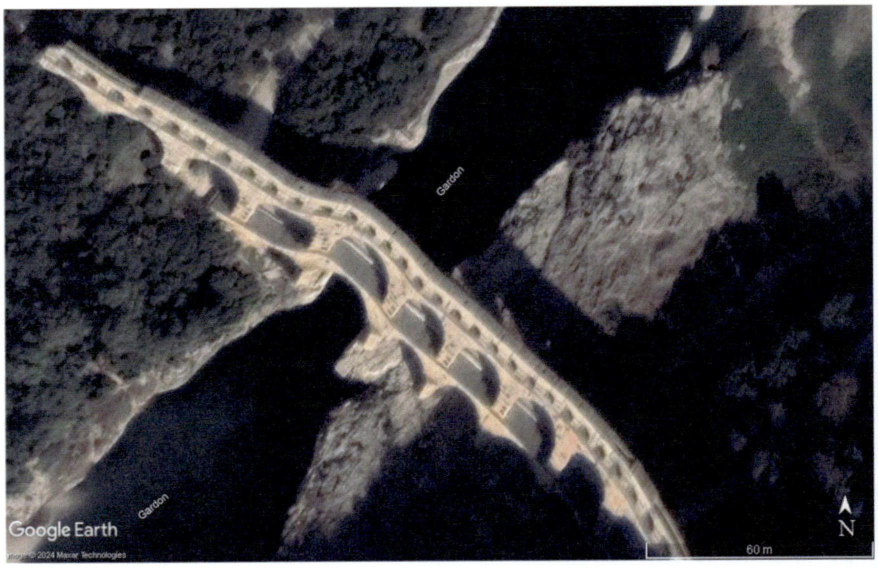

Fig. 2.15 Pont du Gard, southern France. 43°56′50.96″N, 4°32′7.56″E. Scale bar = 60 m, ©Google Earth

Nemausus. The water channel from the Fontaine d'Eure spring source to the *city* descends by only 17 m in height over its entire length, and this is indicative of the precision that Roman water engineers utilised in its design. The Pont du Gard spans the Gardon River, which flows along the bottom of a valley deeply carved in the surrounding plateau. When water reached Nımes, it was stored in a large holding tank called a *castellum*, from which it was piped to all parts of the city. It is a three-storey aqueduct bridge rising to nearly 49 m and made of limestone blocks without the use of mortar. It is a World Heritage Site (https://whc.unesco.org/en/list/344/) (Accessed 4th April 2025).

In some parts of the Middle East, aqueducts were used to transfer water from huge water wheels called *norias*. The noria performs the function of moving water from a lower elevation to a higher elevation, using the energy derived from the flow of a river. It consists of a large, narrow undershot water wheel whose rim is made up of a series of containers or compartments which lift water from the river to an aqueduct at the top of the wheel. The use of norias dates back to Hellenistic and Roman times, but they were considerably developed by the Muslims in North Africa and the Mediterranean lands, including Andalusia. The spectacular Norias of Hama in Syria are a series of 17 wheels along the Orontes River (Miranda 2007). Many of these were functioning at least by the late twelfth century AD. In terms of height, the tallest of the norias is 21 m in diameter (Fig. 2.16).

Fig. 2.16 Top: A Noria and aqueduct at Hama in Syria. 35° 6′53.94″N, 36°47′29.22″E. Scale bar = 40 m, ©Google Earth. Bottom: A noria and aqueduct on the Orontes River in Hama (early twentieth century). (https://timelessmoon.getarchive.net/amp/topics/norias+of+hama) (Accessed 13th April 2025) (Public domain). Other excellent old images of Hama norias are provided in (https://syrianmemories.com/en-gb/blogs/syrian-memories/hama-city-of-waterwheels-norias) (Accessed 13th April 2025)

2.7 Weirs and Fish Traps

One final type of water management that needs discussion is the construction of fish weirs and fish traps in coastal and riverine environments. A famous example of this are the Brewarrina stone weirs on the Barwon River (Fig. 2.17) made by aboriginal inhabitants of Australia (Martin et al. 2023; Daniell and Moggridge 2024). Coastal fish traps have also been constructed in many parts of the world. In Taiwan there are the stone fish weirs of Penghu in Taiwan (Fig. 2.18). Over 500 examples are known, the most famous of which is the Double Heart (Luo et al. 2019). Another example of stone tidal traps are those of Stilbaai in the Western Cape of South Africa (Fig. 2.19). Most of the existing fish traps have been built during the past 300 years, some as recently as the latter part of the twentieth century, but others could date as far back as 3000 years ago (Avery 1975; Kemp et al. 2006). In England, there are many fish weirs in the Bristol Channel (Rippon 2000, p. 222) as on Minehead Beach (51°12′52.56″N, 3°27′16.98″W). Other examples of ancient fish traps are also known from some lake sites, including those from ancient Lake Cahuilla in California (Phukan et al. 2019). Recently, GE has been used to locate four thousand-year-old Mayan fish traps on the waterways of Yucatan in Mexico (Harrison-Buck et al. 2024). There are many examples of ancient fish weirs along rivers in eastern America (Lutins 1992).

Fig. 2.17 Brewarrina, Barwon River, NSW, Australia. 29°57′29.35″S, 146°51′16.43″E. Scale bar = 50 m, ©Google Earth

Fig. 2.18 Penghu Double Heart, Taiwan. 23°13′12.72″N, 119°26′49.69″E. Scale bar = 40 m, ©Google Earth

Fig. 2.19 Stilbaai Fish Traps, Western Cape, South Africa. 34°23′51.65″S, 21°24′45.68″E. Scale bar = 100 m, ©Google Earth

References

Angelakis AN, Antoniou G, Voudouris K, Kazakis N, Dalezios N, Dercas N (2020) History of floods in Greece: causes and measures for protection. Nat Hazards 101:833–852

Angelakis AN et al (2024) Water dams: from ancient to present times and into the future. Water 16(13):1889

Avery G (1975) Discussion on the age and use of tidal fish-traps (visvywers). S Afr Archaeol Bull 30(119/120):105–113

Avni G (2018) Early Islamic irrigated farmsteads and the spread of qanats in Eurasia. Water Hist 10(4):313–338

Barbaix S, De Maeyer P, Chen X, Bourgeois J, Kurban A (2022) An integrated approach to modelling the interaction of the natural landscape and the karez water system in Turpan (Xinjiang, PRC). Landsc Res 47(8):1052–1070

Beckers B, Berking J, Schütt B (2013) Ancient water harvesting methods in the drylands of the Mediterranean and Western Asia. J Anc Stud 2:145–164

Bravard JP et al. (2016) Rise and fall of an Egyptian oasis: Artesian flow, irrigation soils, and historical agricultural development in El-deir, Kharga Depression, Western Desert of Egypt. Geoarchaeology 31(6):467–486

Bruins HJ (2012) Ancient desert agriculture in the Negev and climate-zone boundary changes during average, wet and drought years. J Arid Environ 86:28–42

Brunner U (2006) Notes on gabarbands in Balochistan. Archäol Mitt Iran Turan 38:103–135

Bryan K (1929) Flood-water farming. Geogr Rev 19(3):444–456

Cook H, Williamson T (1999) Water management in the English landscape. Edinburgh Univ Press, Edinburgh, p 273

Daniell KA, Moggridge B (2024) Indigenous water engineering and aquaculture systems in Australia: the Budj Bim cultural landscape and Baiame's Ngunnhu (the Brewarrina aboriginal fish traps). Blue Pap 3(1):18–29

Darby HC (1940) The draining of the Fens. Cambridge Univ Press, Cambridge, p 200

Faiz ME, Ruf T (2010) An introduction to the Khettara in Morocco: two contrasting cases. In: Schneier-Madanes G, Courel MF (eds) Water and sustainability in arid regions. Springer, Dordrecht, pp 151–163

Fletcher R, Pottier C, Evans D, Kummu M (2008) The development of the water management system of Angkor: a provisional model. Bull Indo-Pacific Prehist Ass 28:57–66

Fonder N, Xanthoulis S (2007) Roman aqueduct and hydraulic engineering: case of Nîmes aqueduct and its Pont du Gard Bridge. Water Sci Technol Water Supply 7(1):121–129

Goudie AS (2023) Landscapes of the Anthropocene with Google Earth. Springer, Cham, p 249

Harrison-Buck E, Krause SM, Burg MB, Willis M, Perrotti A, Bailey K (2024) Late Archaic large-scale fisheries in the wetlands of the pre-Columbian Maya Lowlands. Sci Adv 10. https://doi.org/10.1126/sciadv.adq1444

Harvey MD, Flam L (1993) Prehistoric soil and water detention structures (gabarbands) at Phang, Sindh Kohistan, Pakistan: An adaptation to environmental change? Geoarchaeology 8(2):109–126

Hayes-Rich E, Levy J, Hayes-Rich N, Lightfoot D, Gauthier Y (2023) Searching for hidden waters: the effectiveness of remote sensing in assessing the distribution and status of a traditional, earthen irrigation system (khettara) in Morocco. J Archaeol Sci Rep 51:104175

Kahlown MA, Hamilton JR (1996) Sailaba irrigation practices and prospects. Arid Land Res Manage 10(2):179–191

Kemp LV (2006) Ancient stonewall fish traps on the South Coast of South Africa: Documentation, current use, ecological effects and management implications. University of Cape Town, Unpublished thesis. https://open.uct.ac.za/items/05868801-40cc-4f1b-b44d-ab2f4ce1f11a (Accessed 25th August 2025)

Kennedy D, Riley D (1990) Rome's desert frontier from the air. Batsford, London, p 256

Lasaponara R et al. (2016) Puquios: new insights from the integration of remote sensing, GIS-based analyses and geophysical investigations. In: Lasaponara R, Masini M, Orefici G (eds) The ancient Nasca world: new insights from science and archaeology, pp 543–580

Li Q, Guo H, Luo L, Wang X (2022) Automatic mapping of Karez in Turpan Basin based on Google Earth images and the YOLOv5 Model. Remote Sen 14(14):3318

Lightfoot DR (1996) Syrian qanat Romani: history, ecology, abandonment. J Arid Environ 33:321–326

Lightfoot DR (2000) The origin and diffusion of qanats in Arabia: new evidence from the northern and southern peninsula. Geogr J 166(3):215–226

Liu B et al (2017) Earliest hydraulic enterprise in China, 5,100 years ago. Proc Nat Acad Sci 114(52):13637–13642

Luo L et al (2014) Automated extraction of the archaeological tops of qanat shafts from VHR imagery in Google Earth. Remote Sens 6(12):11956–11976

Luo L et al (2019) Low-cost archaeological investigation and rapid mapping of ancient stone tidal weirs in the Penghu archipelago using Google Earth. Sustainability 11(17):4536

Lutins AH (1992) Prehistoric fish weirs in Eastern North America (Doctoral dissertation, Binghamton University, Anthropology Department)

Maghrebi M et al (2022) Anthropogenic decline of ancient, sustainable water systems: qanats. Groundwater 61:139–146

Mamassis N, Moustakas S, Zarkadoulas N (2015) The operation of ancient reclamation works at Lake Copais in Greece. Water Hist 7:271–287

Manley H, Cheetham P, Stewart D, Russell M (2024) A source of confusion: New archaeological evidence for the Dorchester Aqueduct. Britannia. https://doi.org/10.1017/S0068113X24000254

Martin S, Chanson H, Bates B, Keenan-Jones D, Westaway MC (2023) Indigenous fish traps and fish weirs on the Darling (Baaka) River, south-eastern Australia, and their influence on the ecology and morphology of the river and floodplains. Archaeol Oceania 58(1):91–114

McCorriston J, Oches E (2001) Two early Holocene check dams from Southern Arabia. Antiquity 75:675–676

Miranda AD (2007) Water architecture in the lands of Syria: the water-wheels. L'Erma di Bretschneider, Rome, p 370

Ore G, Bruins HJ (2012) Design features of ancient agricultural terrace walls in the Negev Desert: human-made geodiversity. Land Degrad Develop 23:409–418

Ortloff CR (2020) Roman hydraulic engineering: The Pont du Gard aqueduct and Nemausus (Nîmes) Castellum. Water 13(1):54. https://doi.org/10.3390/w13010054

Phukan A, Braje TJ, Rockwell TK, Ullah I (2019) Shorelines in the desert: Mapping fish trap features along the southwest coast of ancient Lake Cahuilla, California. Adv Archaeol Prac 7(4):325–336

Putnam WG (1997) The Dorchester Roman aqueduct. Curr Archaeol 13:364–369

Raikes R (1964) The ancient gabarbands of Baluchistan. East and West 15:26–35

Reddy VR, Reddy MS, Paanisami K (2018) Tank rehabilitation in India: review of experiences and strategies. Agric Water Manag 209:32–43

Remini B, Achour B, Albergel J (2011) Timimoun's foggara (Algeria): an heritage in danger. Arab J Geosci 4:495–506

Rippon S (2000) The transformation of coastal wetlands. Oxford Univ Press, Oxford, p 332

Shaw J, Sutcliffe J, Lloyd-Smith L, Schwenninger JL, Chauhan MS (2007) Ancient irrigation and Buddhist history in Central India: Optically stimulated luminescence dates and pollen sequences from the Sanchi Dams. Asian Perspect 46(1):166–201

Silvester R (1999) Medieval reclamation of marsh and fen. In: Cook H, Williamson T (eds) Water management in the English landscape. Edinburgh Univ Press, Edinburgh, pp 122–140
Tan X, Li Y, Deng J, Wan J, Liu J (2019) The technical history of China's Grand Canal. World Scientific Publishing, Singapore, p 300
Taylor C (1999) Post-medieval drainage of marsh and fen. In: Cook H, Williamson T (eds) Water management in the English landscape. Edinburgh Univ Press, Edinburgh, pp 141–156
Tianduowa Z, Woodson KC, Ertsen MW (2018) Reconstructing ancient Hohokam irrigation systems in the middle Gila River valley, Arizona, United States of America. Human Ecol 46:735–746
Vita-Finzi C (1961) Roman dams in Tripolitania. Antiquity 35:14–20
Wacher J (1974) The towns of Roman Britain. Book Club ssociated, London, p 460
Wilkinson JC (1977) Water and tribal settlement in South-East Arabia: a study of the Aflāj of Oman. Clarendon Press, Oxford, p 276
Williams M (2009) The drainage of the Somerset levels. Cambridge Univ Press, Cambridge, p 320
Woodson MK (2024) Finding and understanding ancient irrigated agricultural fields in southern Arizona. Archeol Pap Am Anthropol Ass 35(1):53–62
Woodson MK, Sandor JA, Strawhacker C, Miles WD (2015) Hohokam canal irrigation and the formation of irragric anthrosols in the Middle Gila River Valley, Arizona, USA. Geoarchaeology 30(4):271–290

3 Landscapes of Agricultural Management

Abstract

With the development of agriculture from the Neolithic onwards, numerous changes have occurred in landscapes. Field systems (e.g. ridge and furrow, lynchets) have been formed, and slopes have been terraced. Crops have been stored in tithe barns and other structures.

Keywords

Celtic fields · Ridge and furrow · Field systems · Terracing · Lynchets · Raised fields · Harvesting

3.1 Introduction

The adoption of agriculture was a global event, with multiple independent origins of farming in Africa, Eurasia, the Americas, and New Guinea, and it increased in spread and intensity throughout the subsequent 10,000 years. In many parts of the world, landscapes show the imprint of past agricultural practices in the shape of field systems and terraces.

3.2 Field Systems

In Britain (https://historicengland.org.uk/images-books/publications/iha-field-systems/heag204-field-systems/) (Accessed 4th April 2025) and various parts of Europe, for example, there are ancient field systems, sometimes called Celtic fields, which date back to prehistoric times. These frequently show up well on air photographs, as is the case with the University of Cambridge collection (https://www.cambridgeairphotos.com/themes/celtic+fields/) (Accessed 4th April 2025). One classic example comes from Grassington in Yorkshire in northern England (Curwen 1928; Joseph 1969) (Fig. 3.1). Grassington lies in the Pennines in the upper valley of the River Wharfe. To the north of the village there are about 100 ha of old field boundaries (Fig. 3.1). These may date to the Iron Age or to Roman times. However, large expanses have been obliterated in recent decades because of extensive ploughing, as is made abundantly clear when one compares Google Earth images of Dorset with earlier photos and maps, such as those of the Royal Commission on Historical Monuments (1972, pp. 118–120; 1975, pp. 117–119).

One consequence of tillage with ploughs is the formation of lynchets and terraces and ridge and furrow patterns in fields. In Roman and Medieval times strip lynchets (*rideaux)* were produced inadvertently by ploughing on steep slopes (Whittington 1962) or, deliberately, to create flat land for cultivation where slopes were otherwise steep. Such features are known from Belgium (Nyssen et al. 2014), France, and Southern England (Dabney and Vieira 2020). Figure 3.2 shows strip lynchets down in Dorset, on the south coast of England.

Ridge and furrow is a pattern of ridges and troughs, sometimes curved, created by a system of repeated ploughing used in Europe during Medieval times and typical of the open-field system that pre-dated enclosures. Surviving ridge and furrow topography is found in Great Britain (Mead 1954), Germany, Ireland, and elsewhere in Europe. In England ridge and furrow is best developed on the clay-rich soils of the Midlands (Buckinghamshire, Warwickshire, Northamptonshire, Leicestershire, Bedfordshire, Oxfordshire, and Western Cambridgeshire) (Hall 1999). The ridges are parallel, ranging from 3 to 20 m apart and up to 61 cm tall. However, they were much taller when in use, and many have been degraded or obliterated by recent ploughing. Indeed, in England (Fig. 3.3), many of the sites of ridge and furrow mapped in the fifties and sixties of the last century (see Harrison et al. 1965) are now no longer visible. Ridge and furrow topography was a result of ploughing with non-reversible ploughs on the same strip of land each year. It is visible on land that was ploughed in the Middle Ages, but which has not been

A. Goudie, *Landscape Heritage of the World with Google Earth*, https://doi.org/10.1007/978-3-031-95555-6_3

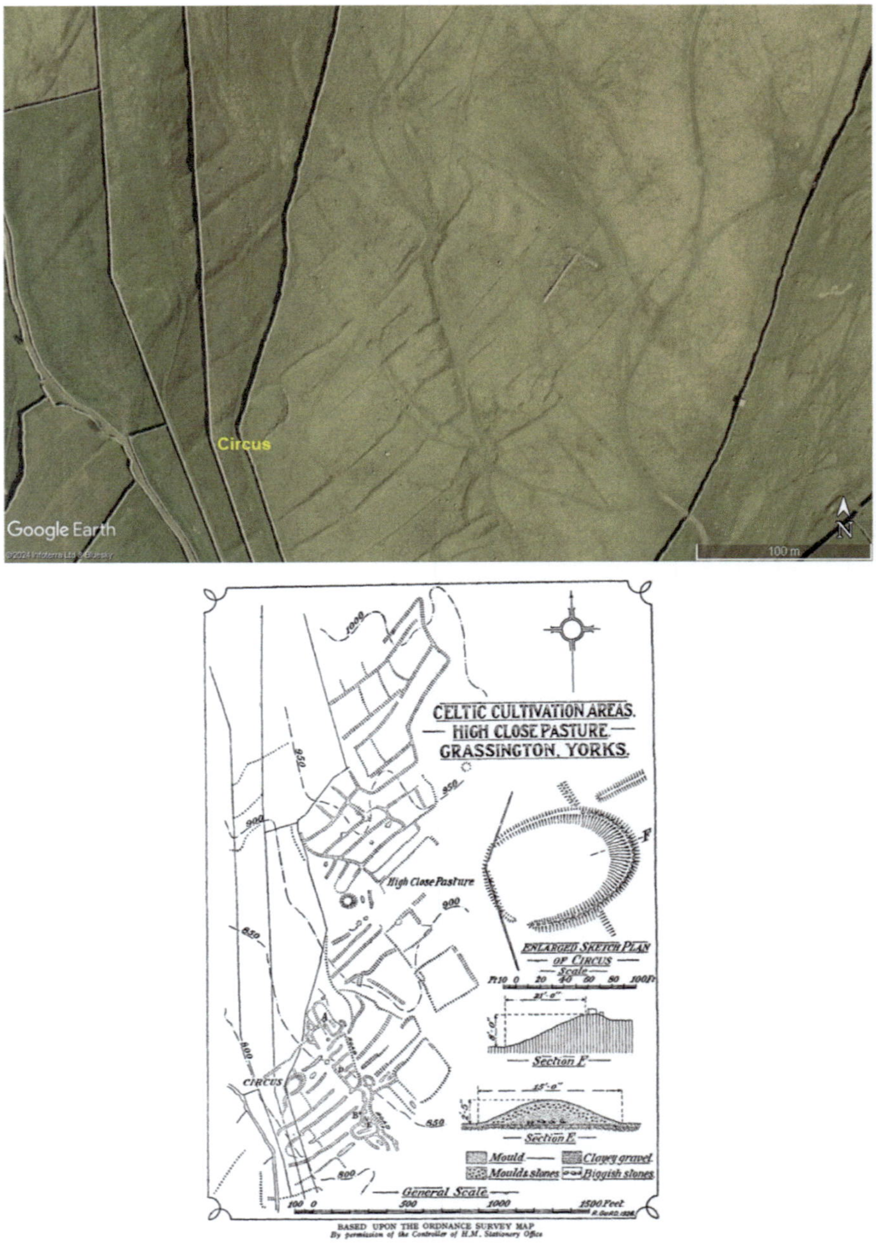

Fig. 3.1 Top. Iron Age/Roman Field Systems at Grassington, Yorkshire, England. 54° 4′50.62″N, 1°59′44.52″W. Scale bar = 100 m, ©Google Earth. Bottom: map from Curwen (1928). The southwestern part of the map shows the area covered by the Google Earth image

ploughed since then. Large common fields were worked in narrow strips of land that were under different ownerships, usually of a furlong (220 yards) in length. At that time, ploughs were pulled by teams of oxen, and they had only one share (the part that cuts into the soil) and a mouldboard that turned the soil over to the right-hand side. The plough was worked across a field in one direction, and the whole team and plough turned at the end and worked back again. The effect of this over time was to turn the soil inwards and build up ridges with deep furrows in between (https://www.bradfordonavonmuseum.co.uk/ridge-furrow) (Accessed 4th April 2025).

The establishment of water meadows, primarily from the sixteenth to the early twentieth centuries, has left a landscape impact in some parts of Europe, including England (Bettey 1999). Water meadows are areas of grassland subject to controlled irrigation by a system of artificial channels (leats) and sluices to increase agricultural productivity on floodplains. They were created as part of agricultural improvement. Irrigation in early spring kept frosts off the ground and so allowed grass to grow several weeks earlier than otherwise, and in dry summers, irrigation kept the grass growing. They were used particularly in the catchments of chalk and limestone streams in Wiltshire, Gloucestershire, Dorset, Berkshire, Hampshire, and Sussex, where clear, nutrient-rich, relatively warm, groundwater-fed streams occurred. They have now largely been abandoned, partly because of a shortage of labour to build, operate, and maintain them, but their impacts

Fig. 3.2 Strip Lynchets, Dorset, England. 50°35′12.42″N, 2° 2′4.15″W. Scale bar = 300 m, ©Google Earth

Fig. 3.3 Ridge and furrow, Northamptonshire, England. 52°17′16.71″N, 1° 3′47.75″W. Scale bar = 90 m, ©Google Earth

can still be seen in the shape of intricate patterns of drainage lines (Fig. 3.4).

In South America, there is abundant evidence for wholesale landscape change being achieved by civilizations such as the Incas (Londoño 2008) and the Mayans (Beach et al. 2023), and then being substantially obliterated after European settlement. In the Mayan Lowlands of Central America, there are raised fields, drainage channels, reservoirs, and other structures that were produced in the 'Mayacene' (Beach et al. 2015, 2023), many of which are now being revealed by LiDAR studies. Raised fields from pre-conquest times, and developed at least in part to provide drainage for wetland soil areas, are indeed still a major feature of large tracts of Amazonia and the Andes (Denevan 2001). Figures 3.5 and 3.6 show examples with varied morphologies from just northwest and southwest of Lake Titicaca in Peru.

3.3 Terraces

Brown et al. (2021 p. 2) have remarked that 'Agricultural terraces are volumetrically the largest and most common landforms that humans have ever produced. They were created on all inhabited continents, and until the nineteenth century, were the only major systematic and worldwide anthropogenic alteration to slopes. Agricultural terraces are therefore probably the most obvious geomorphological marker of the "Anthropocene"…' Brown et al. (2020) pointed out that agricultural terraces are common on sloping lands from New Zealand to Norway, and range in altitude from sea level to over 2,400 m in the French Alps and Peru. Indeed they saw them as the iconic landscapes of the Mediterranean, the Peruvian Andes, the Philippines and the loess of north China. Some are of considerable antiquity, including those of Northumberland in

Fig. 3.4 Top: Water Meadows in the valley of the Frome, near Kingston Maurward, Dorset, England. 50°42′53.69″N, 2°24′12.07″W. Scale bar = 100 m, ©Google Earth. Bottom: Water meadows on the floodplain of the Sherborne Brook in the catchment of the River Windrush, Gloucestershire, England. 51°49′35.80″N, 1°43′20.99″W. Scale bar = 100 m, ©Google Earth. They were in use until the 1930s but have recently been restored by the National Trust (Mayled 1995)

Fig. 3.5 Raised fields, Pomata, Bolivia. 16°15′44.40″S, 69°17′28.57″W. Scale bar = 100 m, ©Google Earth

Fig. 3.6 Raised fields near Lake Titicaca, Bolivia. 15°21′34.15″S, 69°54′51.29″W. Scale bar = 100 m, ©Google Earth

Fig. 3.7 Terracing, Jebel Haraz Region, Yemen. 15° 3′49.89″N, 43°40′1.75″E. Scale bar = 80 m, ©Google Earth

northern England, which date back to the Early to Middle Bronze Age (Brown et al. 2023).

Agricultural cultivation terraces have existed in Europe for at least 8000 years (Brown et al. 2020). Indeed, for many centuries, valley floors, hillside and stream channels have been terraced, notably in the New World's arid and semi-arid highlands (Donkin 1979; Denevan 2001; Whitmore and Turner 2001), around the Mediterranean (Grove and Rackham 2001), in Iberia, on the mountains of South Asia and Arabia, and in sub-Saharan Africa (Grove and Sutton 1989), including northern Ethiopia (Nyssen et al. 2000). In Asia, rice cultivation is often associated with extensive terraced hillsides while some of the Pacific Islands such as Hawaii were terraced for taro cultivation (Allen 1971). In Europe many terraces are associated with vine and olive cultivation (Varotto et al. 2018).

Figure 3.7 shows some terraces from Yemen. These may date back to the third or second millennium BC (Wilkinson 2003, p. 190), while Fig. 3.8 shows some Inca terraces of probable fifteenth-century AD date at Pisac in the Sacred Valley of Peru. The Pisac terraces use granite blocks of varying dimensions. These have been selected and positioned with precision to ensure terrain stability and effective drainage. The combination of worked and natural stones provides structural flexibility and allows for efficient adaptation to the environment (Huaman et al. 2024).

Also in Peru are some extraordinary Inca terraces at Moray near Maras. They are located halfway between Machu Picchu and the city of Cuzco and consist of a group of circular terraces that may have been constructed around the sides of some large sinkholes (Fig. 3.9). The largest is c 30 m deep. Although their purpose may have been agricultural (they were possibly

Fig. 3.8 Pisac Terraces, Peru. 13°24′31.97″S, 71°50′25.32″W. Scale bar = 200 m, ©Google Earth

Fig. 3.9 Circular terraces at Moray, Maras, Peru. 13°19′47.50″S, 72°11′47.77″W. Scale bar = 100 m, ©Google Earth

used to test the effects of different microclimatic conditions on crop growth), their purpose still remains debatable (Earls and Cervantes 2015) (https://hidraulicainca.com/cusco/the-andenes-of-moray-in-cusco/) (Accessed 4th April 2025). It is possible they may have had some ceremonial role. They may have been irrigated, but even here there is uncertainty, for it is conceivable that the water was used for ritual cleansing.

The rice terraces of the Philippines (Fig. 3.10) occur in various mountainous regions of that archipelago, and their construction in the Cordilleras began at a date that is still uncertain. Some authorities suggest 2000 years ago, but others suggest that they were formed after Spanish contact around 400 years ago (Acabado et al. 2019). There is a complex system of communally managed dams, sluices, channels, stone walls, and pipes. They are still viable today and have been declared as a World Heritage Site (https://whc.unesco.org/en/list/722/) (Accessed 4th April 2025). Those of Batad on Luzon Island are particularly admired. Although they are normally termed 'rice terraces', they were, especially in the early days, used for taro cultivation.

In Bali, Indonesia, there is another World Heritage Site (https://whc.unesco.org/en/list/1194/) (Accessed 4th April 2025) called 'Cultural Landscape of Bali Province: the *Subak* System as a Manifestation of the *Tri Hita Karana* Philosophy' that relates to exquisite terrace systems. This site consists of five rice terrace areas and their water temples that cover 19,500 ha. The temples are the focus of a cooperative water management system of canals and weirs, known as *subak,* which dates back to the ninth century. It includes the eighteenth-century Royal Water Temple of Pura

Fig. 3.10 Batad Rice Terraces, Philippines. 16°56′6.53″N, 121° 8′8.95″E. Scale bar = 300 m, ©Google Earth

Fig. 3.11 The Jatiluwih Rice Terraces, Bali, Indonesia. 8°22′18.37″S, 115° 7′53.04″E. Scale bar = 100 m, ©Google Earth

Taman Ayun. The *subak* system of democratic and egalitarian farming practices has enabled the Balinese to become the most prolific rice growers in the archipelago despite the challenge of supporting a dense population. Some of the finest terraces are known as the Jatiluwih rice terraces (Fig. 3.11).

3.3.1 Harvesting

The crops grown in field and on terraces had, once harvested, to be stored, and this also gives rise to some landscape features. Notable amongst these are tithe barns. A tithe barn was a type of timber-framed structure for storing rents and tithes. Farmers were required to give one-tenth (a tithe) of their produce to the established church. From around the twelfth century onwards, barns designed specifically for the storage of tithes were built not just across Britain but also right across northern Europe, particularly in France, Belgium, the Netherlands, and Germany. The total number of surviving medieval barns (dated up to 1550) in Britain may be estimated to be about 200. The barn was controlled by the monks of Beaulieu until Henry VIII dissolved the monasteries, after which it passed into private ownership and then to the National Trust. In Britain, with the passing of the Commutation of Tithes Act of 1836, the system of paying tithes in goods to landlords was replaced by money payments. Local churches, therefore, no longer needed large barns to use for storage. This meant that the tithe barn effectively became redundant.

Great Coxwell Barn in Oxfordshire, England, is a magnificent example of the type (https://www.nationaltrust.org.uk/visit/oxfordshire-buckinghamshire-berkshire/great-coxwell-barn/history-of-great-coxwell-barn) (Accessed 11th

Fig. 3.12 Great Coxwell Tithe Barn, Oxfordshire, England. 51°38′39.84″N, 1°36′45.18″W. Scale bar = 30 m, ©Google Earth

Fig. 3.13 Golghar Granary, Patna, India. 25°37′13.05″N, 85° 8′22.38″E. Scale bar = 40 m, ©Google Earth

April 2025). The barn was built about 1292 for the Cistercian Beaulieu Abbey in Hampshire. It is made of Jurassic Cotswold stone, with rubblestone walls and ashlar buttresses. The barn is 46 m long, 13 m wide, and its roof ridge is 15 m high (Fig. 3.12). Details of this and other Beaulieu-related buildings are provided by Horn and Born (1965). The tithe barn at Abbotsbury in Dorset was even larger—80 m in length.

Places for harvesting crops are known from all agricultural societies, including, for example, the granaries of the ancient Egyptians (as at Luxor) and the Incas (as at Ollantaytambo). The Indus Civilization granary at Mohenjo-Daro (Fig. 7.7) was the largest building that has been unearthed there, and large granaries were also found at Harappa (Fig. 7.6). One interesting but more recent example, from Patna in Bihar, India, is the Golghar Granary. The beehive-shaped structure was designed by John Garstin of the Bengal Engineers, which was part of the East India Company's Bengal Army (Brown 2005). Its construction was completed in 1786 AD with the aim of providing storage against times of famine (Fig. 3.13) (https://architexturez.net/doc/az-cf-177898) (Accessed 12th April 2025).

In the Magreb of North Africa, fortified granaries called *agadirs* were constructed. In Morocco, they are most commonly found in the regions of the High Atlas and Anti-Atlas mountains, and the Draa Valley. Some of them date back to the tenth century. They were often located in rocky, elevated locations to offer protection for surrounding farms and livestock from enemies. In addition to their harvested grains, Amazigh communities used these structures to store all

Fig. 3.14 The Fortified Agadir of Amtoudi, Anti-Atlas Mountains, southern Morocco. 29°14′47.01″N, 9°11′6.32″W. Scale bar = 90 m, ©Google Earth

kinds of valuable belongings. One of the most famous agadirs is that of Amtoudi, perched high up on a hill overlooking the Berber village of that name (Fig. 3.14) (Roble-Marin et al. 2015). It is considered to be both the oldest and most well-preserved agadir in North Africa. This 800-year-old granary is no longer in use.

References

Acabado SB et al (2019) The short history of the Ifugao rice terraces: a local response to the Spanish conquest. J Field Archaeol 44(3):195–214

Allen B (1971) Wet-field taro terraces on Mangaia, Cook Islands. J Polyn Soc 80:371–378

Beach T, Luzzadder-Beach S, Cook D, Turner BL (2023) Progress in physical geography special issue: Physical geography in the Maya Lowlands. Progr Phys Geogr 03091333231162688

Beach T, Luzzadder-Beach S, Krause S, Walling S, Dunning N, Flood J, Guderjan T, Valdez F (2015) 'Mayacene' floodplain and wetland formation in the Rio Bravo Watershed of northwestern Belize. Holocene 25(10):1612–1626

Bettey J (1999) The development of water meadows in the southern counties. In: Cook H, Williamson T (eds) Water management in the english landscape. Edinburgh Univ Press, Edinburgh, pp 179–195

Brown RM (2005) Paṭnā's Golghar and the transformations of colonial discourse. Arch Asian Art 55(1):53–63

Brown A, Walsh K, Fallu D, Cucchiaro S, Tarolli P (2020) European agricultural terraces and lynchets: from archaeological theory to heritage management. World Archaeol 52(4):566–588

Brown AG et al (2021) Ending the Cinderella status of terraces and lynchets in Europe: the geomorphology of agricultural terraces and implications for ecosystem services and climate adaptation. Geomorphology 379:107579

Brown AG et al (2023) Early to middle bronze age agricultural terraces in north-east England: morphology, dating and cultural implications. Antiquity. https://doi.org/10.15184/aqy.2023.1

Curwen E (1928) Ancient cultivations at Grassington, Yorkshire. Antiquity 2(6):168–172

Dabney SM, Vieira DA (2020) Tillage erosion: terrace formation. In: Fath BD, Jorgensen SE (eds) Managing soils and terrestrial systems. CRC Press, Baton Roca, pp 543–550

Denevan WM (2001) Cultivated landscapes of native Amazonia and the Andes. Oxford Univ Press, Oxford, p 426

Donkin RA (1979) Agricultural terracing in the aboriginal new world. Univ Arizona Press, Tucson, p 196

Earls JC, Cervantes G (2015) Inka cosmology in Moray: astronomy, agriculture, and pilgrimage. In: Simada I (ed) The Inka empire: a multidisciplinary approach. Univ of Texas Press, Austin, pp 121–148

Grove AT, Rackham O (2001) The nature of Mediterranean Europe: an ecological history. Yale Univ Press, New Haven and London, p 390

Grove AT, Sutton JEG (1989) Agricultural terracing south of the Sahara. Azania 24:113–122

Hall D (1999) The drainage of arable land in medieval England. In: Cook H, Williamson T (eds) Water management in the English landscape. Edinburgh Univ Press, Edinburgh, pp 28–40

Harrison MJ, Mead WR, Pannett DJ (1965) A midland ridge-and-furrow map. Geogr J 131(3):366–369

Horn W, Born E (1965) The barns of the Abbey of Beaulieu at its granges of Great Coxwell & Beaulieu St. Univ of California Press, Berkeley and Los Angeles, Leonards, p 74

Huaman F et al (2024) Biophysical, spatial, functional, and constructive analysis of the Pre-Hispanic terraces of the ancient city of Pisaq, Cusco, Peru, 2024. Heritage 7(12):6526–6565

Joseph JK (1969) Air Reconnaissance: Recent results, 17. Antiquity 43:220–221

Londoño AC (2008) Pattern and rate of erosion inferred from Inca agricultural terraces in arid southern Peru. Geomorphology 99:13–25

Mayled A (1995) Habitat restoration: the Sherborne Park Estate water-meadows. Biol J Linnean Soc 56(suppl 1):225–227

Mead WR (1954) Ridge and furrow in Buckinghamshire. Geogr J 120(1):34–42

Nyssen J, Poesen J, Haile M, Moeyersons J, Deckers J (2000) Tillage erosion on slopes with soil conservation structures in the Ethiopian highlands. Soil Tillage Res 57(3):115–127

Nyssen J, Debever M, Poesen J, Deckers J (2014) Lynchets in eastern Belgium—a geomorphic feature resulting from non-mechanised crop farming. CATENA 121:164–175

Robles-Marín P et al (2015) Geological risk assessment of Amtoudi Agadir in southern Morocco: a key case for sustainable cultural heritage. Nat Hazards 75:415–440

Royal Commission on Historical Monuments (1972) Historical monuments of the county of Dorset, vol 4, p 142. North. HMSO, London
Royal Commission on Historical Monuments (1975) Historical monuments of the county of Dorset, vol 5, p 155. East. HMSO, London
Varotto M, Bonardi L, Tarolli P (eds) (2018) World terraced landscapes: history, environment, quality of life. Springer, Cham, p 601
Whitmore TM, Turner BL (2001) Cultivated landscapes of Middle America on the eve of conquest. Oxford Univ Press, Oxford, p 311
Whittington G (1962) The distribution of strip lynchets. Trans Inst Brit Geogr 31:115–130
Wilkinson TJ (2003) Archaeological landscapes of the Near East. Univ Arizona Press, Tucson, p 260

4 Landscapes of War

Abstract

Over millennia, a great deal of human ingenuity and effort has been expended on producing defensive structures. These include forts, citadels, castles, moated settlements, causewayed enclosures, trenches, and frontier walls. Special attention is given to hill forts, ringforts, *Nuragi* of Sardinia, *Ksour* of North Africa, causewayed enclosures in the UK, forts in the Roman Empire, Crusader and colonial castles, moated settlements in the UK and Thailand, war trenches, frontier walls, and concentration camps.

Keywords

Forts · Citadels · Castles · Moated settlements · Causewayed enclosures · Trenches · Frontier walls · Concentration camps

4.1 Introduction

Over millennia, a great deal of human ingenuity and effort has been expended on producing defensive structures. These include forts, citadels, castles, moated settlements, causewayed enclosures, trenches, and frontier walls. A detailed discussion of prehistoric defensive landscapes in England and Wales is provided by Coones and Patten (1986), while GE images of some European forts are given in https://theheritagehunter.blogspot.com/2011/06/google-earth-top-10-fortifications.html (Accessed 4th April 2025).

A particularly impressive and long-lived defensive structure is the citadel of the city of Aleppo in Syria (Fig. 4.1). The Citadel was built on a natural limestone outcrop rising some 30 m above the level of the surrounding plain. It is considered to be one of the oldest and largest castles in the world. Usage of the Citadel hill dates back at least to the middle of the third millennium BC. It has been occupied by many civilizations over time—including the Greeks, Armenians, Romans, Byzantines, Ayyubids, Mamluks, and Ottomans. However, the majority of the construction as it stands today is thought to originate from the Ayyubid period. The son of Saladin, Al-Zahir al-Ghazi, ruled Aleppo between 1193 and 1215 AD. It was during this time that the Citadel went through major reconstruction, fortification, and addition of new structures that create the complex of the Citadel as we see it today (Oswald et al. 2014). The great mound is surrounded by a 22 m deep and 30 m wide moat, which dates from the twelfth century. Notable is the fortified gateway, which is accessible through an arched bridge.

Another impressive citadel is the Qaitbay Citadel of Alexandria in Egypt. It was erected on the site of the former Lighthouse, which was one of the Seven Wonders of the Ancient World. The lighthouse, sometimes called the Pharos of Alexandria, was built by the Ptolemaic Kingdom of Egypt, during the reign of Ptolemy II Philadelphus at a strategically vital site. In c1480 AD, the Circassian Mameluke Sultan Al-Ashraf Qaitbay fortified the site as part of his coastal defences against the Turks, who were threatening Egypt at that time. The Citadel was built on, and employed material from, the ruins of the lighthouse, which had been destroyed by earthquakes in the fourteenth century. When Muhammed Ali Pasha became ruler of Egypt in 1805, he renovated the Citadel, restoring and repairing its outer ramparts. He also furnished it with the most modern weapons of the time. The Citadel retained the interest of Mohammed Ali's successors until the year 1882, when the British fleet attacked the city and damaged a large part of it, especially in the area of the Citadel. This attack caused great damage to the fortress. The north and western facades were badly scarred as a result of cannon explosions. The Citadel and the Fortress, though not so much as Aleppo, illustrate how crucial defensive sites are occupied and modified over very extended periods of time (Fig. 4.2).

A. Goudie, *Landscape Heritage of the World with Google Earth*, https://doi.org/10.1007/978-3-031-95555-6_4

Fig. 4.1 Top: Aleppo Citadel, Syria. 36°11′57.47″ N, 37° 9′47.53″ E. Scale bar = 100 m, ©Google Earth Bottom: Fortified entrance to Aleppo Citadel in 1965, showing the arched bridge (*Photo* A.S. Goudie)

The repeated use of defensive sites in strategically important locations is no better displayed than on Malta. The fortifications of this island consist of a number of walled cities (e.g. Mdina), citadels, towers, forts, batteries, redoubts, entrenchments, and pillboxes (https://en.wikipedia.org/wiki/List_of_fortifications_in_Malta) (Accessed 4th April 2025). These were constructed over hundreds of years, from around 1450 BC to the Second World War, and they are a result of Malta's strategic position in the Mediterranean and its desirable natural harbours consisting of flooded river valleys (rias). The fortifications of Malta were greatly developed while it was ruled by the Order of St John between 1530 and 1798, so that by the close of the eighteenth century, Malta had extensive fortifications around both the Grand Harbour of Valletta and Marsamxett, as well as a coastal defence system consisting of towers, batteries, redoubts and entrenchments. Following a brief occupation by the French, the islands fell under British rule in 1800. Between the 1870s and the 1900s the British built a number of polygonal forts and batteries. In the 1930s Fort Campbell was constructed, and was one of the last fortifications to be built in Malta.

An excellent example of a fort built by the Order of St John, is Fort Manoel (Fig. 4.3). It was created in the eighteenth century. The first stone was laid in 1723, and the fort was an active military establishment by 1734. Fort Manoel is built in the shape of a square, with a pentagonal bastion on each corner, giving it the shape of a star fort. The fort saw use again during the Second World War, when a battery of anti-aircraft guns was deployed there. During the war it suffered considerale damage as a result of German and Italian aerial bombing. It was eventually decommissioned in 1964.

Fig. 4.2 Top: The Citadel and Lighthouse in Alexandria, Egypt. 31°12′49.66″ N, 29°53′9.37″ E. Scale bar = 100 m, ©Google Earth. Bottom: The Citadel in the 1780s (https://commons.wikimedia.org/wiki/File:Louis-Fran%C3%A7ois_Cassas,_Le_Grand_Pharillon_du_Port_D%27Alexandrie.jpg) (Accessed 4th April 2025)

4.2 Hill Forts

Hill forts are earthworks used as a fortified refuge or defended settlement (Hawkes 1931). There are more than 3000 in the British Isles (https://historicengland.org.uk/images-books/publications/iha-hillforts/) (Accessed 8th June 2023), many datng back to the Bronze and Iron Ages, and they are especially well developed in central-southern Britain encompassing Wessex (Wiltshire, Dorset, and Hampshire) and the South Downs (wholly within Sussex), and extending northwards and westwards to the Cotswolds and the Welsh borders. Reasons for their adoption and subsequent abandonment were discussed by Cunliffe (1990, 1994).

In spite of their name, not all are on hills, and not all were necessarily built primarily for defence (Lock 2022). Some were built on the sites of Neolithic causewayed enclosures. In Europe they date to the Bronze and Iron Ages, though some were used in the post-Roman period and as sites of castles in medieval times.

They consist of one or more ramparts and ditches. Maiden Castle and Badbury Rings in Dorset, southern England, are notable examples. However, hillforts of broadly similar age are known from other parts of Europe, including the Iberian Peninsula. Some of those in the Ukraine, dating back to the seventh century BC, are so big that they have been termed 'mega-hillforts' and they can cover up to 8000 ha (Daragan 2020). Hillforts are known from outside Europe, and occur, for example, in Fiji (Nunn et al. 2021).

Covering over 45 ha, and with some ramparts rising to more than 6 m, Maiden Castle, Dorset (Fig. 4.4), is the largest hill fort in Britain. It lies on a chalk knoll and has excellent views

Fig. 4.3 Fort Manoel, Malta. 35°54′10.47″ N, 14°30′20.27″ E. Scale bar = 100 m, ©Google Earth

Fig. 4.4 Top: Maiden Castle, Dorset, England. 50°41′42.15″ N, 2°28′11.30″ W. Scale bar = 300 m, ©Google Earth. Bottom: The western entrance to Maiden Castle, Dorset, England (*Photo* A.S. Goudie)

Fig. 4.5 Top: Badbury Rings, Dorset, England. 50°49′34.51″ N, 2°3′7.53″ W. Scale bar = 200 m, ©Google Earth, Bottom: Badbury Rings showing the three ramparts and one of the two main entrances (*Photo* A.S. Goudie)

across the surrounding countryside. The first occupation dates back to 3800 BC, and there are traces of a Neolithic causewayed camp. However, the first major fortifications were built in the Iron Age around 600 BC. Subsequently in the late fourth century AD, a temple and ancillary buildings were constructed on the site by the Romans. The entrance on the western side is complex and well preserved (Fig. 4.5). It was complicated, so it was harder to get into the hill fort, and the western entrance had a 200 m long winding corridor. Maiden Castle, like Badbury Rings, was probably an important centre of the Durotriges, one of the Celtic tribes living in southern England prior to the Roman conquest. Full details of the site and of the excavations undertaken by Rick Wheeler in the 1930s are given by the Royal Commission on Historical Monuments (England) (1970).

Another important hillfort in Dorset is Badbury Rings (Fig. 4.5). It occupies a low summit 3 km north of the River Stour (Papworth 2011). This 16.5 ha late Iron Age feature has three circuits of massive ramparts (https://heritagerecords.nationaltrust.org.uk/HBSMR/MonRecord.aspx?uid=MNA137102) (Accessed 4th April 2025) but there are also traces of Bronze Age round barrows, and in the northwest of the image, of a Roman road. The multivallate hillfort dates from c 150 BC, was built by the Durotriges, and was probably stormed by the Romans. In the Roman era a temple was located immediately west of the fort, and there was a Romano-British town known as *Vindocladia* a short distance to the southwest.

Although traces of hut circles have been found at both Maiden Castle and Badbury Rings, those from Tre'r Ceiri on the Llŷn Peninsula in western Wales are especially striking (Abram 2022, pp. 258–259) (Fig. 4.6). The outlines of no less than c 150 huts survive on the summit plateau and are in superb condition. It has a circumference of c 620 m and an area of 2.5 ha, and the walled defences consist of a main inner wall and a partial outer wall on the less naturally defensible

Fig. 4.6 Tre'r Ceiri, Hillfort and Hut Circles, Wales. 52°58′28.63″ N, 4°25′25.95″ W. Scale bar = 100 m, ©Google Earth

side. The inner wall stands in places up to 4m high, including parapet breastwork, and Darvill (1996, p. 116) suggested that 'this wall is probably the most completely surviving defensive work at any hillfort in Britain'. Evidence suggests this Late Iron Age hillfort was first built around 200 BC, though most of the archaeological finds date from AD 150–400, showing that the site continued as a settlement during the subsequent Roman occupation (https://medievalheritage.eu/en/main-page/heritage/wales/trer-ceiri-hillfort/) (Accessed 4th April 2025).

On the South Downs in Sussex, Cissbury Ring is an Iron Age hillfort, and is the second largest in Britain. It covers c 24 ha (Fig. 4.7). The earthworks that form it have a jelly bean shape and were built around the beginning of the Middle Iron Age (possibly around 400 BC) but it was reinforced at the end of the Roman occupation to defend against invading Saxons. Cissbury is a univallate fort, consisting of a hilltop enclosure with a single rampart accompanied by a ditch and a low counterscarp bank. Cissbury Hill also has numerous Neolithic flint mines, which are marked by depressions, particularly at the southwest end (http://sussexarch.org.uk/saaf/cissbury.html) (Accessed 4th April 2025). The many hillforts of south-east England are mapped and described by Hamilton and Manley (2001).

In Wiltshire stands Old Sarum, the ruined and deserted site of the earliest settlement of Salisbury. It too is an Iron Age hillfort, erected around 400 BC, and controlling the intersection of two trade paths and the Hampshire Avon. It has impressive univallate earthworks. However, it is also notable for medieval features, including the ground plan of a

Fig. 4.7 Cissbury Ring Hillfort, Sussex, England. 50°51′37.40″ N, 0°22′58.14″ W. Scale bar = 200 m, ©Google Earth

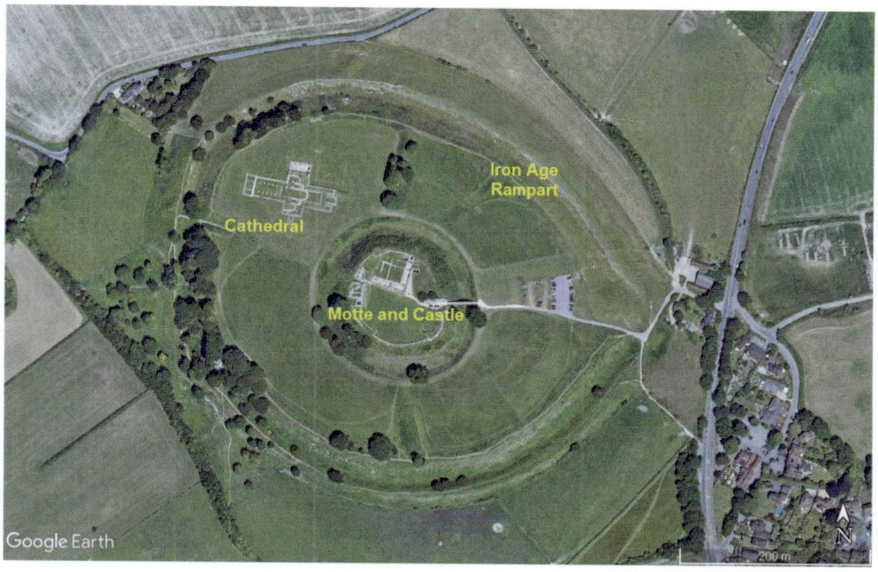

Fig. 4.8 Old Sarum, Wiltshire, England. 51°5′36.12″ N, 1°48′14.87″ W. Scale bar = 200 m, ©Google Earth

cathedral and a Norman motte and bailey castle in the centre, built by William the Conqueror (Fig. 4.8). Motte and bailey was a common form of castle in the eleventh and twelfth centuries in Britain. Mottes were mounds, either natural or man-made, which were surmounted by a tower. Around them was a ditch, sometimes filled with water. For man-made mottes the earth needed to build it was taken from the surrounding ditch to create another defensive feature. The bailey was an outer enclosure, which was also defended by a fence and sometimes a ditch, providing a first, outer line of defence for the motte (https://www.castlesandmanorhouses.com/types_03_motte.htm) (Accessed 8th April 2025). A royal palace was constructed within this castle for King Henry 1, and was subsequently used by Plantagenet monarchs. The see of Old Sarum was transferred to New Sarum (Salisbury) in 1219, and the cathedral was finally razed in 1331 (Coones and Patten 1986, pp. 200–202).

4.3 Ringforts

Ringforts are small circular fortified settlements, typically 30–50 m in diameter, which were built from the Bronze and Iron Ages up to about the year 1000 AD. They are widely distributed in northern Europe, and especially in Ireland, where over 40,000 sites have been identified. These may have been dispersed farmsteads and may have been used as enclosures for livestock. With their banks and ditches they seem to have had a defensive function as a deterrent to cattle-raiding (Limbert 1996). The term is not without controversy, and Fitzpatrick (2009, p. 303) has written, 'The wide range of native enclosures that antiquaries and field archaeologists working in Ireland since the nineteenth century recognised, the fact that many such enclosures are not forts in the true sense of that word, and that their chronology in terms of occupation potentially spans a millennium, is evidence enough that the term "ring-fort" is discredited as a meaningful definition of a native settlement form'.

Rathra (Fig. 4.9) is located near Castlerea, County Roscommon, and is an enclosure, thought to have been built sometime during the Late Bronze Age/Early Iron Age. It has a series of four ditches and earthen banks surrounding a central area. The outer bank and ditch enclose an area about 145 m in diameter with 3 larger inner banks (up to 3 m in height), surrounding a raised central area roughly 97 m in diameter. Within this central area are two barrows associated with Bronze Age burials.

The Grianan of Aileach (Fig. 4.10) is another Irish ringfort and lies atop the 244 m high Greenan Mountain at Inishowen in County Donegal. The wall is about 4.5 m thick and 5 m high, and the feature has an average internal diameter of c 23 m. It is encircled by three enclosures and is therefore termed trivallate. There was substantial restoration in the 1870s. Parts of it may date back to the Iron Age, but most of it was probably constructed during the eighth or ninth century AD as the seat of Cenél nEógain, rulers of the ancient kingdom of Aileach (https://www.archaeology.ie/monument-of-the-month/archive/grianan-of-aileach-co-donegal) (Accessed 4th April 2025).

In the northwest of the Iberian Peninsula there are many fortified settlements called *castros* (Cunliffe 2001, Chap. 8). They date from the late Bronze Age well into Roman times. They occur in hilltop situations, have defensive walls, and circular houses. A notable example is that on the Atlantic coast of Galicia at Castro de Baroña (Fig. 4.11). Fortified settlement towers also occur in the United Arab Emirates (Kennet 1992).

Fig. 4.9 Rathra Ringfort, Ireland. 53°45′21.49″ N, 8°25′18.56″ W. Scale bar = 70 m, ©Google Earth

Fig. 4.10 Grianan of Aileach, Ireland. 55° 1′25.63″ N, 7°25′39.57″ W. Scale bar = 40 m, ©Google Earth

Fig. 4.11 Castro de Baroña, Spain. 42°41′40.32″ N, 9° 1′55.08″ W. Scale bar = 50 m, ©Google Earth

4.4 Nuragi

In the Bronze Age, during the late second millennium BC, defensive structures were developed on Sardinia, which are unique to that Italian island. They are known as *nuragi* and they consist of circular truncated cones with corbel-vaulted internal chambers. They were constructed of local stone (Mariani et al. 2024). They might also have had some religious significance. The best example of these is at Su Nuraxia di Barumini (Fig. 4.12), a World Heritage Site (https://whc.unesco.org/en/list/833/) (Accessed 4th April 2025). At this impressive but rather threatening site there is a massive central tower of three chambers connected by a spiral staircase, originally over 18.5 m high. There are also some perimeter towers and a defensive wall. They are mainly made of dark Sardinian basalt. During some time in the seventh century BC the site was sacked by the Carthaginians, but it was inhabited until the third century AD. A village of small roundhouses had developed around the tower, particularly on the eastern side. Some 6,500 *nuraghi* survive to the present day in Sardinia, with an overall average density of 0.27 per km^2 rising in some districts to 0.9 km^2 (Lilliu 1959).

Fig. 4.12 Top: Su Nuraxi, Sardinia, Italy. 39°42′21.52″ N, 8°59′27.36″ E. Scale bar = 80 m, ©Google Earth. Middle: Round houses. Bottom: Tower (*Photos* A.S. Goudie)

4.5 Ksour and Caravanserais

In the Maghreb of North Africa there are numerous fortified settlements called *ksour* (singular *ksar*). Many of these castles were constructed in oasis regions of the northern Sahara of mud brick by Berber people and were used as granaries, caravanserais, bases for trans-Saharan trade, and the like. The Ksar Draa in Algeria is a circular structure located in a sea of dunes and shows up particularly well on GE (Fig. 4.13). There are two outer walls and the remains of a number of internal buildings.

The Ksour of Mauritania, founded in the eleventh and twelfth centuries, are a World Heritage Site (https://whc.unesco.org/en/list/750) (Accessed 4th April 2025), as is the Ksar of Ait-Ben-Haddou in pre-Saharan Morocco (https://whc.unesco.org/en/list/444) (Accessed 4th April 2025).

Caravanserais often showed defensive features. They provided secure lodging for travellers, merchants, and caravans and served as way stations where merchants and travellers could safely stop and rest along the way, and obtain water. A notable example is the Qasr al-Hayr al-Sharqi in central Syria. It was built of mud brick and stone by the Umayyad caliph Hishan ibn Abd al-Malik in 728–729 AD and was ultimately abandoned at the start of the fourteenth century. It is believed to be the oldest identified example of an Islamic caravanserai and is on the World Heritage tentative list (https://whc.unesco.org/en/tentativelists/1298/) (Accessed 8th April 2025) (Fig. 4.14). One of the so-called 'desert castles', it lies c 110km to the northeast of Palmyra in the Syrian steppe lands and is situated near the crossroads of routes linking the caravan city of Palmyra (see Fig. 1.3) to Rusafa (see Fig. 4.18) and Raqqa on the upper course of the

Fig. 4.13 Ksar Draa, Algeria. 29°32′5.40″ N, 0°11′0.77″ E. Scale bar = 40 m, ©Google Earth

Fig. 4.14 Two enclosures of Qasr al-Hayr al-Sharqi, Syria. 35°4′26.61″ N, 39°4′17.59″ E. Scale bar = 1000 m, ©Google Earth

Syrian Euphrates, and to Dayr al-Zawr, the Jazira and lower Mesopotamia (Genequand 2005). The Syrian Desert was an important link between Occident and Orient.

4.6 Roman Forts

The Romans were great fort builders and their forts are located from the northern parts of their lands (e.g. along Hadrian's Wall) to those in the hearts of the deserts of Egypt and Jordan.

One of the latter, El-Deir, occurs in the Kharga Oasis, Egypt (Fig. 4.15). This is located in the Western Desert, and was at an important intersection connecting caravan roads from the Darfur province of Sudan (ancient Nubia) to the Nile Valley, a journey of c 1700 km. The huge enclosure measures 73 m^2 and had twelve round towers interspersed along its thick mudbrick walls. The monument is very well preserved, with walls up to 15 m high and with remains of 12 round towers. However, it has been subject to some damage and looting (Abdelmohsen and Ismail 2025). Its dimensions suggest that an important garrison could have been stationed there. It is thought to date to the second or third century AD (Tallet et al. 2013).

Rather similar in form to El-Deir, though smaller, is the fort of Qasr Bshir (Mobene) in the heart of the Jordanian Desert (Clark 1987) (Fig. 4.16). It is almost square in plan, measuring about 57 m by 54 m, which is small by Roman fort standards. Its four imposing corner towers still stand three storeys—that is, 13 m—high. The masonry used to build this *Castellum* consists of some enormous stones of megalithic proportions (https://www.world-archaeology.com/features/qasr-bshir/) (Accessed 4th April 2025). It was constructed on the southeastern fringe of the Empire during the reign of the Emperor Diocletian and his colleagues

Fig. 4.15 Top: El-Deir Fort, Kharga, Egypt. 25°35′48.08″ N, 30°43′49.88″ E. Scale bar = 50 m, ©Google Earth. Bottom: Ground Shot (*Photo* A.S. Goudie)

Fig. 4.16 Roman Fort of Qasr Bshir, Jordan. 31°20′14.22″ N, 35°58′52.35″ E. Scale bar = 40 m, ©Google Earth

between 293 and 305 AD, and is on a World Heritage Site Tentative List (https://whc.unesco.org/en/tentativelists/1553/) (Accessed 4th April 2025). It belongs to the chain of forts and watchtowers that is known as the *Limes Arabicus* and was meant to protect the province of Arabia against roaming desert nomads.

Portchester Castle in Hampshire (Fig. 4.17), southern England, is a Roman fort situated at the head of Portsmouth Harbour. It was concerned with naval strategy and with the command of the harbour, the Solent, the Isle of Wight, and Southampton Water. It is what is known as a Saxon Shore fort, and it was constructed by Carausius (Frere and St Joseph 1983, pp. 79–81), the man appointed by the Emperor Diocletian to command the Roman fleet in the Channel in about AD 285. These Saxon Shore forts were built over the course of the third century to meet the threat presented by Saxon pirates who were then raiding the south coast of Roman Britain (https://www.english-heritage.org.uk/visit/places/portchester-castle/history-and-stories/history/) (Accessed 4th April 2025). It was occupied during the Saxon period (Cunliffe 1970) and then in the late eleventh century a Norman tower keep castle was built within the shore fort walls in the northwest corner. Another Saxon Shore fort, which shows up well on GE, is Pevensey Castle in East Sussex (50°49′9.74″ N 50°49′9.74″ N, 0°20′4.00″ E).

Resafa (Fig. 4.18) is an archaeological site in Syria (Hof 2016), and it was known in the Byzantine era as Sergiopolis. It was a city located in the Roman province of Euphratensis and is situated southwest of the city of Raqqa and the River Euphrates. The site dates to the ninth century BC, when a military camp was built by the Assyrians. During Roman times, it was a desert outpost fortified to defend against the Sasanian Empire, and it flourished as it was located on the caravan routes linking Aleppo and Palmyra. Ramparts and

Fig. 4.17 Portchester Roman Saxon Shore Castle, Hampshire, England. 50°50′13.83″ N, 1°6′49.09″ W. Scale bar = 100 m, ©Google Earth

Fig. 4.18 Resafa, Syria. 35°37′45.61″ N, 38°45′28.98″ E. Scale bar = 300 m, ©Google Earth

buildings were erected there by Emperor Justinian in the sixth-century AD. The well-preserved walls of Resafa are over 450 m in length and about 300 m in width; round or square towers were erected regularly around its perimeter. Resafa, in a dry region, had no spring or running water, so it depended on large cisterns to capture the winter and spring rains. In the fourth century AD it became a pilgrimage town for Christians coming to venerate Saint Sergius, one of the earliest Christian martyrs. He had been decapitated in that town, along with a fellow soldier, Bacchus, by Emperor Maximian for professing Christianity. The city of Resafa, which became a bishop's see, preserved his relics in a fortified basilica, and it was improved by Emperor Justinian and became one of the greatest pilgrimage centres in the East.

4.7 Forts of Rajasthan and Sind

In the northwest of the Indian sub-continent lie the states of Rajasthan (India) and Sind (Pakistan). For centuries it was the scene of conflict between princely states and with Afghan and Mughal invaders, and so large number of magnificent forts were constructed, mostly on commanding hills. Indeed, the majority of the cities of this arid area have forts, including Jaipur, Jaisalmer, Bikaner, Jodhpur, and Jalore. Many of these forts were located on important trade routes and often served as clan capitals of the Rajput warrior class of this region. They were largely built by the leaders of the various princely states that dominated the area from the eighth to the eighteenth centuries AD. Some of them are now on the UNESCO World Heritage List (https://whc.unesco.org/en/list/247/) (Accessed 4th April 2025). Jain (2022) discusses the nature and distribution of forts in India as a whole.

One example is the fort that dominates the city of Jaisalmer in northwest Rajasthan (Fig. 4.19). It is the second oldest fort in Rajasthan, and was built in 1156 AD by the Rawal (ruler) Jaisal, from whom it derives its name. It has massive yellow Jurassic limestone walls. The fort is 460 m long and 230 m wide and is built on a hill that rises 76 m above the surrounding countryside. It has 99 bastions. During medieval times, Jaisalmer benefitted from its strategic location on the camel trade route between India and Central Asia. The people of Jaisalmer had trading connections with Afghanistan, Iran, and other Central Asian towns such as Tashkent, Khorasan, and Samarkand. Many people still live within the bounds of the fort, but it has now lost most of its trading and defensive role. This is because of a number of factors, including the advent of British rule, the emergence of maritime trade, the growth of the port of Bombay, and the partition of Indian after Independence.

The city of Jaipur is blessed with a whole suite of hill forts that crown the area's sandstone hills, including Amber, Jaigarh (Fig. 4.20), and Nahargarh. Jaigarh was built by Maharaja Jai Singh in 1726, one of the state's greatest rulers.

At Kot Diji In Pakistan there is a slightly younger but major fort with a perimeter of 1.8 km, a sinuous outline, and whuch was built of bricks by Mir Sohrab Khan Talpur, between 1785 and 1795 AD (Fig. 4.21). Ali (2011) describes many of the other forts of Pakistan.

4.8 Miscellaneous European and Eurasian Forts

Saxon defensive ramparts built in England included those at Wareham in Dorset and Wallingford in Oxfordshire (Coones and Patten 1986, pp. 141–144). But after this, when William

Fig. 4.19 Top: Jaisalmer Fort, Rajasthan, India. 26°54′45.91″ N, 70°54′45.67″ E. Scale bar = 200 m, ©Google Earth

Fig. 4.20 Top: Part of Jaigarh Fort, Jaipur, Rajasthan, India. 26°59′13.16″ N, 75°51′27.74″ E. Scale bar = 100 m, ©Google Earth. Bottom: Bejeweled Maharaja Jai Singh II of Jaipur. (https://commons.wikimedia.org/wiki/File:Jai_Singh_II.jpg) (Accessed 4th April 2025)

Fig. 4.21 Kot Diji Fort, Pakistan. 27°20′43.81″ N, 68°42′26.38″ E. Scale bar = 200 m, ©Google Earth

the Conqueror conquered England in 1066, he set about building castles in order to exert his control. One of these was Corfe Castle in Dorset, England (Fig. 4.22) (National Trust 2015). It was built upon a natural mound and controlled the route from Swanage to Wareham through the Purbeck Hills. The keep at Corfe was built in the early twelfth century for King Henry I, William the Conqueror's son. Standing 21 m tall and positioned atop a 55 m high hill, this gleaming tower of Purbeck limestone could be seen from miles around. King John, who reigned from 1199 to 1216 AD, liked Corfe Castle and wanted more luxury during his frequent visits. He therefore built the gloriette (a pocket palace) in c 1201-5 in the shadow of the old Norman keep (Reeve and Thurlby 2005). In the 1640s, England was in the grip of civil war and Corfe Castle, owned then by the Royalist Bankes family, found itself on the front line of conflict between Parliament and King Charles I. After six centuries of keeping enemies at bay, the Royalist Governor of the castle surrendered and in 1646 the House of Commons ordered its demolition. That is why it is in ruins today.

Castles were also built by the English to control the Welsh. In 1295 AD King Edward I, having conquered Wales, commissioned the building of Beaumaris Castle on Anglesey (Fig. 4.23). Its construction was designed and directed by James of St. George (c. 1230–1309), King Edward I of England's chief architect. This was part of a chain of fortifications around the North Wales coast. Others include Conwy, Caernarfon, and Harlech (Kenyon 2018). These four castles are on the World Heritage List (https://whc.unesco.org/en/list/374/) (Accessed 4th April 2025). The castle was built on a marsh, and that is where it found its name; the Norman-French builders called it *beau marais* (fair marsh). It was never completed, and building stopped in 1330. In architectural terms Beaumaris is often regarded as the most technically perfect castle in Britain. Its ingenious and perfectly symmetrical concentric 'walls within walls' design, involving no less than four successive lines of fortifications and a moat, was state of the art for the late thirteenth century. Supply ships could sail right up to the castle dock.

Across the Irish Sea, and of much later date, is the Charles Fort on Kinsale Harbour, County Cork, and Ireland (Fig. 4.24). Kinsale was the most utilised natural harbour in Ireland in the sixteenth and eighteenth centuries when England was fighting both Spain and France for dominion over the seas. The fort is of the pentagonal bastion type with one section of the outer wall built in star fashion. This design enabled it to counter any attack from the sea. Some of the outer defences are 16 m high. It is named after King Charles second. It was built between 1677 and 1682, and remained in use as a British Army barracks for two hundred years afterwards, before they relinquished it following the Anglo-Irish Treaty of 1921. Shortly after, anti-Treaty forces destroyed the barracks and burned the buildings.

A major phase of fort building in southern England was that of the so-called Palmerston Forts. These were built as a result of a decision in Lord Palmerston's premiership to deter the French from attacking naval facilities following the recommendations of a Royal Commission on the Defence of the United Kingdom which had reported in 1860. Many of these, especially in the vicinity of Plymouth and Portsmouth, show up well on GE (see, for instance, Fort Brockhurst near Portsmouth, 50°48′53.48″ N, 1° 9′14.34″ W). This was completed in 1862.

Once cannons became introduced into warfare, forts were designed with a star shape, also known as bastion forts, to maximise defensive capabilities against cannon fire by creating interlocking fields of fire and minimising dead zones,

Fig. 4.22 Top: Corfe Castle, Dorset, England. 50°38′23.95″ N, 2°3′32.19″ W. Scale bar = 80 m, ©Google Earth. Bottom: Corfe Castle from the southeast (*Photo*: A.S. Goudie)

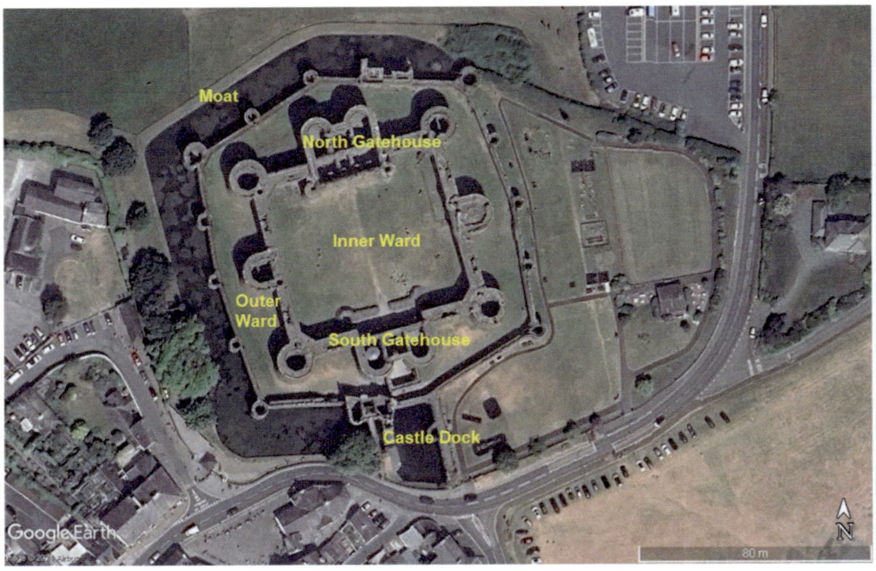

Fig. 4.23 Beaumaris Castle, Wales. 53°15′52.98″ N, 4°5′21.67″ W. Scale bar = 80 m, ©Google Earth

Fig. 4.24 Charles Fort, Kinsale, Ireland. 51°41′48.02″ N, 8°29′55.88″ W. Scale bar = 100 m, ©Google Earth

Fig. 4.25 Castelo de Almeida, Portugal. 40°43′32.80″ N, 6°54′18.32″ W. Scale bar = 80 m, ©Google Earth

making it difficult for attackers to find cover (https://www.castlesandmanorhouses.com/types_10_star.htm) (Accessed 8th April 2025). This morphology was first seen in Italy in the mid-fifteenth century. One European fort with star-shaped elements is that of Almeida in Portugal (Fig. 4.25). It was constructed in this region due to its significant strategic importance in close proximity to the border between Portugal and Spain. Following John IV of Portugal's restoration of the Crown of Portugal in 1640, the monarch sought to consolidate and strengthen his throne. He decided to build a new fort at Almeida. The fortress seen today is arranged on an irregular hexagonal plan and consists of six polygonal, irregular, and uneven-sized bastions which are connected by curtain walls and forming a perimeter of 2.4 km. It consists of a complex system of bulwarks, ravelins (triangular fortifications or detached outworks, located in front of the curtain walls and bastions of a fortress), platforms, trenches, and a moat. Star-shaped forts are a common morphological type, and there are many fine examples, not only in Europe, but also in North America and Japan. They show up especially well on satellite images (https://www.theatlantic.com/photo/2023/11/photos-star-forts/675936/) (Accessed 4th April 2025) and (https://www.euspaceimaging.com/blog/2023/12/01/star-forts-in-satellite-images/) (Accessed 4th April 2025).

Yet another fine example of a star-shaped fort is that of Palmanova in northern Italy (Fig. 4.26). This was included as a World Heritage Site (Venetian Works of Defence between the sixteenth and seventeenth centuries) (https://whc.unesco.org/en/list/1533/) (Accessed 4th April 2025). Using all the latest military innovations of the sixteenth century, this small town, built by the Venetian Republic, partly as a defence against the Turks, was a fortress in the shape of a nine-pointed star.

Fig. 4.26 Palmanova, Italy. 40°43′32.80″ N, 6°54′18.32″ W. Scale bar = 900 m, ©Google Earth

Between the points of the star, ramparts protruded so that the points could defend each other. A moat also surrounded the town, and three large, guarded gates allowed entry. The construction of the first circle, starting in 1593, with a total circumference of 7 km, took 30 years. A second phase of construction took place between 1658 and 1690, and the outer line of fortifications was completed between 1806 and 1813 when the area was under Napoleonic domination.

The Venetians were great fortress builders around the shores of the Mediterranean, not least in Greece. One castle of note is the Palamidi Fortress in Nafplio, Peloponnese. Similarly, the Fortezza in Rethymno, Crete, stands as an imposing emblem of the Venetian era. Construction of the Fortezza began in 1573, and it was completed by 1580. It was captured by the Ottomans in 1646. The Fortezza has an irregular plan, and its walls have a total length of over 1.3 km (Fig. 4.27).

In Sweden, the Hanseatic town of Visby (57°38′20.23″ N, 18°18′4.91″ E) has extensive thirteenth-century ramparts and is a World Heritage Site (https://whc.unesco.org/en/list/731/) (Accessed 4th April 2025).

Por Bazhyn (Por Bajin), in the far southeast of Siberia, is an enigmatic site which consists of a gigantic clay complex (~35,000 m^2) built by the Uyghurs in the eighth century that completely covers an island in Lake Tere-Khol (Fig. 4.28). It is situated close to the northern margins of the so-called Uyghur Khaganate, an empire that, at one point, encompassed the whole of modern-day Mongolia and parts of southern Siberia. Radiocarbon dates by Kuitems et al. (2020) show that construction of the site started in 777 AD. The site is rectangular in shape, with dimensions of 215 × 160 m. The massive external rammed earth walls are 8–10 m high and about 10 m thick at their base (Alfimov 2013). The interior is

Fig. 4.27 The Fortezza of Rethymno, Crete. 35°22′19.86″ N, 24°28′17.32″ E. Scale bar = 100 m, ©Google Earth

Fig. 4.28 Por Bazhyn, Russia. 50°36′54.66″ N, 97°23′5.94″ E. Scale bar = 100 m, ©Google Earth

taken up by two large yards, a central building complex, and a chain of small yards along the northern, western, and southern walls. It has many characteristics of Chinese Tang architecture of the time and has been interpreted as a summer palace built by Khagan Bögü, which, after damage to the palace by an earthquake and Khagan's conversion to Manichaeism, was converted into a Manichaean monastery. Following his death in 780 and the abolition of Manichaeism afterwards, the monastery was abandoned.

The city of Derbent, in Dagestan, Russia, is located on the western Caspian Sea coast, in a location which forms a narrow gateway between the Caspian Sea and the Caucasus Mountains. Derbent played an important role in the development of the Silk Roads; it was located on the Caspian trail, which was of strategic importance, connecting Europe and Western Asia. Derbent claims to be one of the oldest cities in the world and the southernmost one of the Russian Federation. It is believed to be 5,000 years old. It underwent great development from the fifth century BC, and the fortress was built by the Sasanians in order to defend Western Asia from various nomadic and steppe tribes (https://en.unesco.org/silkroad/content/derbent) (Accessed 4th April 2025). Construction of the current citadel (Fig. 4.29), fortified walls, and city walls began c. 570 AD. Derbent was also an important religious centre (https://unesco.ru/en/news/52-derbent/) (Accessed 4th April 2025).

4.9 Crusader and Other Castles

Particularly in Syria and Jordan there are some remarkable castles that were built by the Crusaders (Kennedy 2001; Al Haddad et al. 2023). The Crusades were a series of religious

Fig. 4.29 Derbent Citadel, Dagestan, Russia. 42° 3′10.39″ N, 48°16′27.83″ E. Scale bar = 100 m, ©Google Earth

wars between Christianity and Islam. The best-known of these military expeditions were those to the Holy Land in the period between 1095 and 1291. They had the objective of reconquering Jerusalem and its environs from Muslim rule. The Crusaders built castles, which were a tangible symbol of the dominance of a Latin Christian minority over a largely hostile majority population, and also acted as centres of administration. Many castles were constructed (see https://en.wikipedia.org/wiki/List_of_Crusader_castles) (Accessed 4th April 2025). Three are discussed here: Kerak and Shobak in Jordan, and Krak des Chevaliers in Syria.

Construction of Kerak (Fig. 4.30 top) began in the 1140s on the site of a more ancient city, and it took approximately twenty years to complete. The castle occupies a steep-sided plateau, surrounded by deeply incised wadis, which provide some measure of natural defence. The fortress was called the 'Crow's Castle' because it was situated atop a narrow ridge that towered threateningly over the King's Highway, which was the main land route between Damascus in the north and Egypt and Arabia to the south. Because of its position east of the Dead Sea, it was able to control Bedouin herders as well as the trade routes from Damascus to Egypt and

Fig. 4.30 Crusader Castles. Top: Kerak Castle, Jordan. 31°10′50.73″ N, 25°9′47.14″ E. Scale bar = 100 m, ©Google Earth. Middle: Shobak Castle, Jordan. 30°31′53.12″ N, 35°33′40.49″ E. Scale bar = 100m, ©Google Earth. Bottom: Krak Des Chevaliers, Syria. 34°45′24.64″ N, 36°17′41.45″ E. Scale bar = 200 m, ©Google Earth

Mecca. However, in 1188 it fell to Muslim armies (Anderson 2017).

Shobak Castle (also called Montreal) (Fig. 4.30 middle) was built in 1115 by Baldwin I with the dual purpose of extending the area of Frankish settlement east of the Dead Sea and of controlling the desert road, by which caravans moved between Syria in the north and the Red Sea and Egypt in the south (https://whc.unesco.org/en/tentativelists/1552/) (Accessed 4th April 2025). However, like Kerak, it was short-lived as a Christian fortress, for it fell to Muslim forces in May 1189. They added to its fortifications. Shobak is strategically located on the summit of a cone-shaped hill. The castle is oval in shape and has tall masonry walls and a series of rectangular and semi-circular towers.

The site of Krak des Chevaliers (Fortress of the Knights) (Fig. 4.30 bottom) in Syria, was taken over by the Hospitaller Order of Saint John of Jerusalem in the 1140s and fortifications were created (https://whc.unesco.org/en/list/1229/) (Accessed 4th April 2025) After a second phase of building was undertaken in the thirteenth century, Krak des Chevaliers became a concentric castle, and this phase created the outer wall and gave the castle its current appearance. The final Hospitaller version of the fortress had concentric walls consisting of an outer circuit wall with built-in round projecting towers and an inner wall, also with towers. The whole structure was built on a natural platform of solid rock. At its peak, Krak des Chevaliers housed a garrison of around 2,000, but in 1271 the Mamluk Sultanate captured it. The castle commanded two crucial communication routes between the coast and the Muslim cities of Hama and Homs on the Orontes plain.

4.10 Colonial Castles

The European colonial powers built castles to defend their empires. One of these is the castle at Cape Coast in Ghana (Fig. 4.31). It has been used and occupied by a number of different nations over the years (Hyland 1995). It was originally a Portuguese trading post, established in 1555, which was named *Cabo Corso*. In 1653, a timber fort was constructed by the Swedish Africa Company, who named it Carolusborg after their king. The fort exchanged hands several times, being conquered by Danes, then by the Dutch, and from 1664 by the British, who over the next century and more turned the by-then stone fort into a much larger and more powerful castle, well stocked with cannons, which reached its present appearance by about 1795. Colonial rivalry between England and France peaked in 1757 during the Seven Years' War. A French naval squadron bombarded Cape Coast Castle, leaving it badly damaged, and after 1760, the English reconstructed it. The fort was used to harbour enslaved Africans before they were loaded onto ships and sold in the Americas, especially the Caribbean. At any one time up to 1,500 Africans were held in the castle dungeons waiting for the next slaving ship. It is now a World Heritage Site (https://whc.unesco.org/en/list/34/) (Accessed 4th April 2025). Cape Coast Castle is one of three slave castles on the coast of Ghana. The others are at Elmina and Christiansborg. A list of the many European colonial establishments on the coast of West Africa is https://commons.wikimedia.org/wiki/Forts_and_factories_on_the_(West_African)_Gold_Coast_to_1872 (Accessed 4th April 2025).

Fig. 4.31 Cape Coast Castle, Ghana. 5° 6′13.36″ N, 1°14′27.43″ W. Scale bar = 60 m, ©Google Earth

Fig. 4.32 Top: Fort Jesus Portuguese Fort, Mombasa, Kenya. 4°3′45.64″ S, 39°40′47.69″ E. Scale bar = 80 m, ©Google Earth. Middle and Bottom: Fort Jesus (*Photos* A.S. Goudie)

Another African fort from the European colonial era is Fort Jesus at Mombasa on the coast of Kenya (Kirkman 1974) (Fig. 4.32). Like Cape Coast it has had a chequered history that has involved two colonial powers—Portugal and Britain—and Omani Arabs. Fort Jesus was captured and recaptured at least nine times between 1631 and 1895. It was built between 1593 and 1596 by order of King Felipe II to guard the island of Mombasa. It was the only fort maintained by the Portuguese on the Swahili Coast and is a testament to the first successful attempt by a Western colonial power to establish its influence over trade in the Indian Ocean and to provide a protected stepping stone to its colonies in India. It has imposing 18 m walls and five bastions. The morphology of the fort represents the rough outline of a person lying on their back, with their head towards the sea. On the landward side great ditches were excavated in the coral rock to create a dry moat designed to render uninvited access extremely difficult. The fort was also designed to be resistant to cannon fire. It is a World Heritage Site (https://whc.unesco.org/en/list/1295/) (Accessed 4th April 2025). Other Portuguese forts showing up on GE are discussed in https://www.colonialvoyage.com/category/colonial-forts-on-google-earth/ (Accessed 4th April

Fig. 4.33 Forte Real de São Filipe Santiago, Cape Verde. 14°54′57.22″ N, 23°36′7.05″ W. Scale bar = 70 m, ©Google Earth

2025). They include the sixteenth-century fort of San Sebastian on Mozambique Island, which is located at 15°1′42.87″ S, 40°44′37.73″ E. It also is a World Heritage Site (https://whc.unesco.org/en/list/599/) (Accessed 4th April 2025).

Another Portuguese fortress, on the other side of the African continent, is Forte Real de São Filipe, Santiago, in the Cape Verde Islands (Fig. 4.33). This is a sixteenth-century building in the city of Cidade Velha. This was the first permanent European settlement in the tropics and the city's historic centre is a World Heritage Site (https://whc.unesco.org/en/list/1310) (Accessed 4th April 2025). The fort was part of a system of defence for the city, which also included six smaller forts on the coast and a wall along the port. The fortress consists of three bastions towards the interior of the island, and towards the north and south, it is protected by a cliff. The fortress was built under Philip I of Portugal between 1587 and 1593, which was a little after Sir Francis Drake had raided Santiago in 1585. It was a key port of call, being an exceptionally important centre on the routes for international maritime trade, including those between Africa and the Cape of Good Hope, Brazil, and the Caribbean. In 1712 the French arrived on the island of Santiago and sacked the town. Thereafter, the Portuguese moved Santiago Island's centre of power a few km east to the town of Praia, and Forte Reale slid into decay. El Jadida in Morocco (33°15′23.84″ N, 8°30′8.45″ W) is another Portuguese Fort that shows up well on GE.

The Portuguese also constructed forts in the Gulf Region, in such places as Bahrain, Iran and Oman. The northern coastal cities of Oman were under Portuguese rule between 1507 and 1656. Two impressive fortresses are those of Jalali (São João) and Al Mirani, both built by the Portuguese around 1580. Muscat harbour was strongly defended against attack from the sea by these twin forts, by the Muttrah fort further to the west (also built by the Portuguese in the late sixteenth century) and by other fortifications on the rocky ridges surrounding the bay (Fig. 4.34).

Nor surprisingly given the size of her overseas empire, Spain also constructed a series of forts in the Americas and the Philippines, and some of these are World Heritage Sites, as in Panama (https://whc.unesco.org/en/list/135/gallery/) (Accessed 4th April 2025). In Acapulco, Mexico, is the star-shaped Fort of San Diego (Fig. 4.35). It was one of the most important Spanish fortifications along the Pacific Coast of Mexico and its exceptional harbour was a major base for Spanish galleons (Schurz 1918). These left Acapulco in March or April arriving in June or July in Manila to avoid the storm season in Southeast Asia. They returned in July or August, taking advantage of favourable winds that allowed them to sail east across the Pacific arriving at Cape Mendocino in California, then turning south towards Acapulco. The fort was completed in 1617 to the designs of the Dutch military engineer Adrián Boot (Sluiter 1949). It was extensively damaged by an earthquake in 1776, and so was demolished and rebuilt to designs of Ramón Panón. Construction of the new fort began in 1778 and was completed in 1783. It housed 2,000 soldiers with provisions and ammunition for a year. Fort San Diego fell to insurgent forces in 1813 during Mexico's War of Independence after a siege lasting several months. King Ferdinand II decommissioned the fort in 1816.

In southwest Sri Lanka the Dutch created some great fortifications at Galle (Fig. 4.36). It is a World Heritage Site and the citation stated: 'Founded in the sixteenth century by the Portuguese as a fortified town, Galle reached the height of its development in the eighteenth century under Dutch colonial rule. It is the best representation of a fortified city built by the Europeans in South Asia (https://whc.unesco.org/en/list/451/) (Accessed 4th April 2025). The fort was built first in 1588 by the Portuguese, but the Dutch captured the fort in

Fig. 4.34 Portuguese Al Mirani Fort in Muscat, Sultanate of Oman. Top: Google Earth image 23°37′1.77″ N, 58°35′37.12″ E. Scale bar = 60 m, ©Google Earth. Bottom: Ground image of Muttrah fort rising above the harbour and the modern city (*Photo* A.S. Goudie)

Fig. 4.35 Fort of San Diego, Acapulco, Mexico. 16°51′0.48″ N, 99°54′8.22″ W. Scale bar = 90 m, ©Google Earth

Fig. 4.36 Galle Fort, Sri Lanka. 6°1′38.07″ N, 80°13′2.92″ E. Scale bar = 400 m, ©Google Earth

1640 and carried out fortifications up until the early eighteenth century. They decided fully to encircle the entire peninsula on which Galle lies by building impregnable fortifications as a defence against other colonial agencies in the region (Jinadasa 2020). They built 13 bastions with coral and granite stones over an area of 52 ha. The British took over the fort on 23 February 1796, and the cricket stadium immediately to the north of the fort's main gate is part of their legacy. Galle still has various Dutch buildings, including the old hospital.

Both in North America, Indo-China and North Africa, the French constructed forts to protect their colonies. In North America there was a star-shaped fort called Fort Ticonderoga (Fig. 4.37) that was built between 1755 and 1757 in northern New York State near the south end of Lake Champlain. The fort was of strategic importance during the eighteenth-century colonial conflicts between Great Britain and France, and again played an important role during the American Revolutionary War.

In the Sahara Desert, forts were constructed as a defence against the Tuareg and other threats. One example is Fort Flatters (Fig. 4.38), Bordj Omar Driss, which was built in the east of Algeria. It was named after Paul Flatters, a French soldier who spent a long period as a military administrator in that country. He was the leader of the Flatters expedition, an ill-fated attempt to explore the route of a proposed Trans-Saharan railway from Algeria to the Sudan. Almost all members of the expedition were massacred by hostile Tuaregs in 1881. The Fort was built in 1904, expanded in 1918, and was used by the French Foreign Legion.

The German colonial empire was relatively modest in size and short-lived. One example of a German colonial fort is that at Namutoni in the Etosha region of Namibia (formerly

Fig. 4.37 Fort Ticonderoga, USA. 43°50′30.37″ N, 73°23′15.35″ W. Scale bar = 90 m, ©Google Earth

Fig. 4.38 Fort Flatters, Algeria. 28°6′15.09″ N, 6°48′41.63″ E. Scale bar = 60 m, ©Google Earth

German southwest Africa) (18°48′29.16″ S, 16°56′23.69″ E). It was built in 1896 and ceased to be used only 16 years later. It is now a tourist resort. In the centre of the erstwhile German capital at Windhoek is the Alte Feste (22°34′9.91″ S, 22°34′9.91″ S).

4.11 Moated Settlements

In many parts of the world there are settlements which are surrounded by moats and which therefore are called 'moated settlements'. They occur in the British Isles, where many of them are of medieval age (Emery 1962). More than 8500 have been found in England alone (Coveney 2015). Figure 4.39 shows Broughton Castle, a medieval fortified manor house in Oxfordshire, England. Such moats may well have played a defensive role, but they may have had other functions, including indicating status and providing a source of fish.

This is the case with the many moated settlements found in Thailand (Yang et al. 2024), where they date to the first millennium AD and consist of settlements surrounded by deep ditches and, sometimes, ramparts. These ditches may have been several metres deep and 25 to 40 m wide (Mudar 1999). Several potential functions for the earthworks have been theorised. Explanations have included not only defence, but also aquaculture, flood protection, symbolism or ritual, and water storage for domestic consumption as well as providing some mitigation for drought conditions (Scott and O'Reilly 2017). One example is the Ban Non Wat Moated Settlement (Fig. 4.40). GE has been used to map the distribution of moated sites in the area and has identified many new sites (O'Reilly and Scott 2015). Moated settlements are

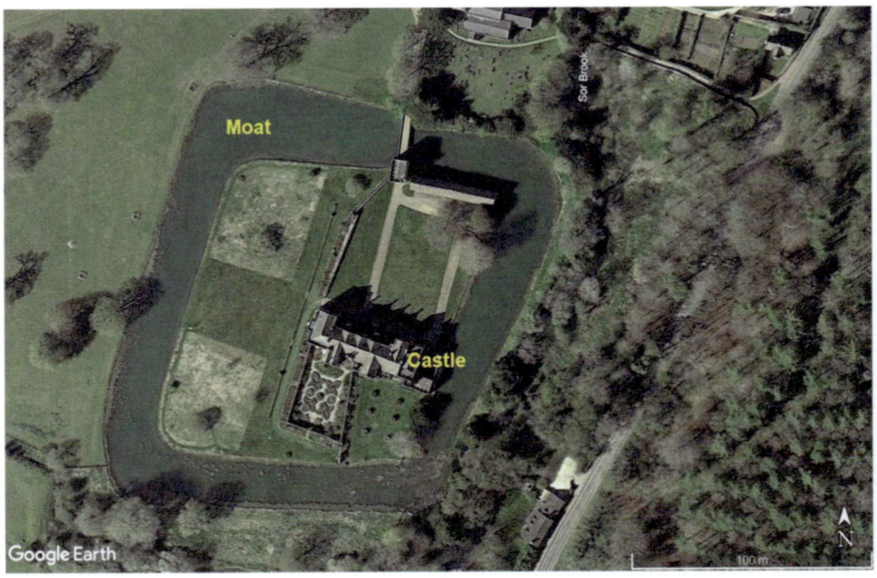

Fig. 4.39 Broughton Castle Moated Settlement, Oxfordshire, England. 52°2′26.41″ N, 1°23′29.28″ W. Scale bar = 100 m, ©Google Earth

Fig. 4.40 Ban Non Wat Moated Settlement, Thailand. 15°16′3.56″ N, 102°16′35.75″ E. Scale bar = 200 m, ©Google Earth

widespread in other parts of Asia, including Japan and China, where some examples date back to as much as 8000 years (Altaweel and Zhuang 2018).

4.12 War Trenches and Emplacements

In the twentieth century, trench warfare became a major aspect of military activity. One location where this was the case, and where GE has been used to plot the earthworks, is Ma'an in southern Jordan (Winterburn 2021) (Fig. 4.41). These defensive structures were built by the Turks. Ottoman Ma'an was a major communications point, the seat of a provincial district governor, and a fortified military base on the Hejaz Railway in southern Jordan. Blessed with plentiful water, it had long been a centre on the pilgrimage caravan route to Mecca. When the railway reached the town in 1904, a large station complex was established 1.5 km to the south, and Ma'an became the 'Gateway to Arabia'. The Hejaz Railway was a 1302 km long, single-track, narrow-gauge railway linking Damascus and Medina. Construction of the 'holy' railway began in 1900, and this monumental feat of engineering and a triumph of Islamic financing was completed as far as Medina by 1908. The Ottoman loss of Aqaba to the Arab armies and T. E. Lawrence led to the arrival in Maᶜan, in August, of the German military advisor General Erich von Falkenhayn to supervise a defence plan and fortify the garrison station by surrounding it with trenches. The conflict landscapes around Ma'an Station are the most comprehensive and best-preserved First World War landscapes in Jordan. The extensive trench systems represent

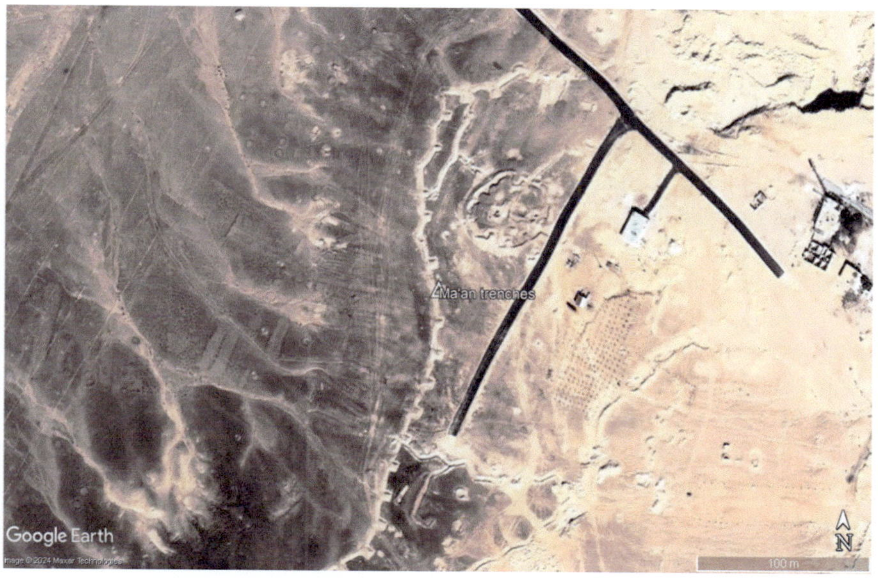

Fig. 4.41 Turkish First World War defences at Ma'an, Jordan. 30°10′54.33″ N, 35°45′25.61″ E (see Saunders and Faulkner 2010). Scale bar = 100 m, ©Google Earth

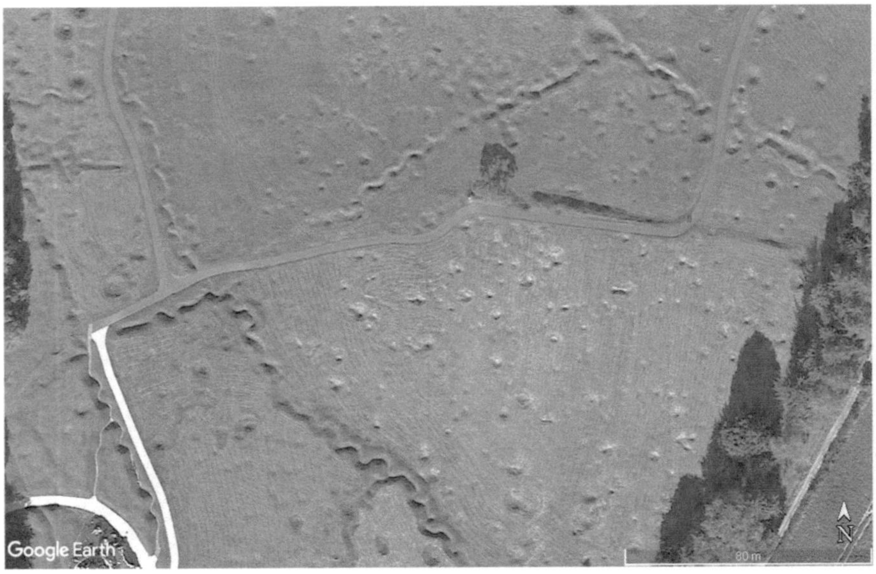

Fig. 4.42 First World War Trenches and Craters at Beaumont Hamel, France. 50°4′28.75″ N, 2°38′59.20″ E. Note also the multitude of craters. Scale bar = 80 m, ©Google Earth

Falkenhayn's attempt to transform an unprotected garrison and centre of railway engineering into a fortress. His plan was successful, and 'fortress Maᶜan' survived an attack by the Northern Arab Army in April 1918 (Saunders and Faulkner 2010).

The First World War battlefields of Flanders were serrated by lines of trenches. Figure 4.42 shows some trenches and craters from a battlefield at Beaumont Hamel in the Somme Department of northern France.

Other important landscapes associated with past wars, are gun emplacements, but space precludes they can be dealt with here, except for one example, the High Angle Battery, that was built in 1892 to 1896 to protect the great new naval harbour at Portland, southern England (Fig. 4.43) (https://navalhistoria.com/the-verne-battery/) (Accessed 4th April 2025). Having been constructed within a disused quarry, the battery's guns were hidden from enemy view. The 'high angle' that the 9 in. 12 tonne guns fired at ensured that their great shells dropped down from above to inflict maximum damage on the vulnerable upper decks of an enemy battleship. The battery initially had positions for six guns. Shells were stored in two magazines, while between these were a troop shelter, laboratory, and store buildings. A small railway was used to carry shells around the site, and there are extensive underground tunnels. With the invention of smaller craft like torpedo boats and with advancements in ship speed, the guns were less likely to score a hit, resulting in the battery's decommissioning in 1906. The battery went out of use after only 10 years, never firing a single shot in combat.

On the other side of the English Channel in Normandy, France, is Pointe du Hoc (Fig. 4.44). This points out into the English Channel and became of huge significance in World War II when it was involved with the greatest amphibious military exercise of all time—the D Day Landings of 1944. The point was strategically located as the highest point between Utah Beach to the west and Omaha Beach to the east. The Germans, who occupied France, had installed a series of large circular gun emplacements and machine gun posts during 1943 and 1944. These posed a threat to Allied landings, and so were heavily bombed from the air and shelled from the sea, creating a large number of craters. During the American assault on Omaha and Utah beaches on 6 June 1944, Army Rangers scaled the 30 m high cliffs and in a daring raid seized the German artillery pieces that could have fired on the landing American troops (Zaloga 2011).

4.13 Frontier Walls

In addition to the forts, castles, citadels, etc. discussed so far, there are various frontier walls that have been constructed, including the Great Wall of China (Fig. 4.45) (Jing 2015). This was built from the seventh century onwards across the historical northern borders of ancient Chinese states and Imperial China as protection against raiding nomadic groups from the Eurasian Steppe, and to mark the limits of civilization. Most of the wall we see today was built by the Ming Dynasty (AD 1368–1644). Winding along the crests of mountainous terrain, it is c 6 m wide at its base and between 6 and 9 m in height. The defensive capabilities of the Great Wall were enhanced by the construction of some 20,000 watchtowers, troop barracks, garrison stations, and signalling beacons. It is not a single 'great wall'. The frontier walls built by different dynasties have multiple courses. Collectively, they stretch from Liaodong in the east to Lop Nor in the west, and from the present-day Chinese-Russian border in the north to the Tao River in the south. In all this spans over 21,000 km.

Fig. 4.43 High Angle Battery, Portland, England. Top: 50°33′28.78″ N, 2°25′59.89″ W. Scale bar = 50 m, ©Google Earth. The six recessed gun emplacements are on the east side. Bottom: Ground shot from south (*Photo* Heather Viles)

In northwest China, there are groups of forts that are located by ancient sections of the Great Wall. These include Jiayuguan and those of the Yumen Pass. At Jiayuguan (Fig. 4.46) the western sector of the Great Wall was began by General Feng Sheng, a Ming Dynasty officer, in 1372. This section of the wall is mainly built with rammed earth. There is a two-storey watchtower on each corner. It has a bailey with three storied buildings, inner and outer walls, two gatehouses, a moat, and beacon towers. The bailey has a perimeter of 640 m and an area of 25,000 m^2. The whole construction lasted 168 years from 1372 to 1540. Located on the Silk Road, it served as the western door to China and marked the extent of Ming imperial power (Rovero and Tonietti 2014).

The Yumen Pass (Yumenguang) fort, to the west of Dunhuang, was founded during the Han Dynasty (206 BC–220 AD) and served as a strategic military fortification on the North Silk Road. It was constructed with rammed earth (Fig. 4.47). The site is an ancient fortress of the Han Great Wall. It was comprised of four walls, c 25–26 m long, is 4–5 m wide at its base, and has a height of around 9.6 m. Important conservation work is being undertaken by the Dunhuang Academy (Wang et al. 2010).

There are impressive defensive walls in other parts of the world, including Iran (Nokandeh et al. 2006) with its Great Wall of Gorgan (Fig. 4.48). This is 195 km long and 6–10 m wide and features over 30 fortresses spaced at intervals of between 10 and 50 km. It is surpassed only by the wall

Fig. 4.44 The Gun Emplacements and Craters at Pointe du Hoc, Normandy, France. 49°23′46.18″ N, 0°59′19.88″ W. Scale bar = 100 m, ©Google Earth

systems of the Great Wall of China as the longest defensive wall in the world. It was built from c 420–530 AD (https://whc.unesco.org/en/tentativelists/6199/) (Accessed 8th June 2023). Another major Iranian wall evident from GE is that of Gawri (Alibaigi 2019).

Significant fortified walls were also constructed by the Romans. The *Limes Germanicus* were built in Germany and they used either a natural boundary such as a river or they employed an earth bank and ditch with a wooden palisade and watchtowers at intervals. They were associated with a system of linked forts. Running from the northwest of the country to the Danube in the southeast, the total length was 568 km, and it included at least 60 forts and 900 watchtowers. In Romania were the Dacian Limes, stretching over 100 km.

Castellum Fectio fort is in the Netherlands (Fig. 4.49). This was a *castellum* in the province of Germania Inferior and was established in 4 or 5 AD. It was located at the place where the River Vecht (*Fectio*) branched off from the Rhine. It was built by order of Emperor Tiberius. Later, under Emperor Claudius, Fectio became part of the *Limes Germanicus*. The archeological site contains the remains of a fort, port, cemetery, and a civilian settlement. Around 200 AD, it seems that river siltation prevented access to the boats. It was thus abandoned in 275 AD and was never rebuilt though it has recently been largely reconstructed. At its height, the *Limes Germanicus* stretched from the North Sea outlet of the River Rhine to near Regensburg on the River Danube. It is part of a World Heritage Site (https://whc.unesco.org/en/list/430/) (Accessed 4th April 2025).

In Hungary, the Csörsz Trench and ramparts were built in the Great Hungarian Plain. Other notable Roman defensive walls occurred in Britannia: the 118-km-long Hadrian's Wall in northern England; and the Antonine Wall, a 60-km-long fortification in Scotland which was started by Emperor Antonius Pius in 142 AD as a defence against the 'barbarians' of the north (https://www.worldheritagesite.org/list/Frontiers+of+the+Roman+Empire) (Accessed 8th June 2023).

Hadrian's Wall (Fig. 4.50) was begun in AD 122 in the reign of the Emperor Hadrian (Goldsworthy 2022). Most of it took six years to build. Running from Wallsend-on-Tyne in the east to Bowness-on-Solway in the west it was a stone or turf wall with large ditches in front and behind, stretching across the whole width of the island. Soldiers were garrisoned along the line of the wall in large forts, smaller milecastles (fortlets), and intervening turrets. Forts, each housed around 600 men. One well-preserved fort is Housesteads. In Roman times this was covered in buildings including a headquarters building, granaries, a commandant's house, hospital, barrack, latrines, and a bathhouse. Hadrian's Wall was declared a World Heritage Site in 1987, and in 2005 it became part of the transnational 'Frontiers of the Roman Empire' World Heritage Site, which also includes sites in Germany (https://www.english-heritage.org.uk/visit/places/hadrians-wall/hadrians-wall-history-and-stories/history/) (Accessed 4th April 2025).

The Roman walls at Lugo in northwest Spain (Fig. 4.51) were built in the later part of the third century AD to defend the Roman town of Lucus against local tribesmen and Germanic invaders. Although ravaged by fire and invasions, the walls are remarkable for their preservation and are beautifully seen in the centre of the modern town. They are described in the World Heritage citation as 'the most complete and best-preserved example of Roman military architecture in the Western Roman Empire' (https://whc.unesco.org/en/list/987/) (Accessed 4th April 2025). Of the original towers 46 have survived, and there are a further 39 that are wholly or in part dismantled. There are also ten gates, five of which are Roman. The walls have a total length of over 2 km and are in the shape of an oblong rectangle. This occupies an area

Fig. 4.45 Top: Great Wall of China. 40°21′6.19″ N, 116°0′30.73″ E. Scale bar = 700 m, ©Google Earth. Bottom: Great Wall near Beijing (*Photo* A.S.Goudie)

of nearly 1.7 ha. Their height is between 8 and 10 m, and they have a width of 4.2 m, reaching 7 m at some specific points. The walls are built of stone facings of slate with some granite, with a core filling of an artificial conglomerate of slate slabs and worked stone pieces from Roman buildings, interlocked with lime mortar.

4.14 Concentration Camps and Political Prisons

One of the most harrowing consequences of war was the construction and operation of concentration camps. The term *concentration camp* was first used to refer to those which were set up by the Spanish in Cuba during the Ten Years' War (1868–1878). It was also applied to camps set up by the United States during the Philippine-American War. The British set up concentration camps in South Africa during the Boer War (1899–1902) for interning Boers, while the Germans established concentration camps during the Herero and Nama genocide in southwest Africa (1904–1907) at places like Aus and Shark Island.

The Nazi regime in Germany built and used a huge number of camps, and at its peak, the Nazi concentration camp system had as many as 15,000 camps. One of the most notorious of these was Buchenwald in Weimar (Fig. 4.52). It was established in 1937, and its inmates came from all over Europe and the Soviet Union, and included Jews, Poles, other Slavs,

Fig. 4.46 Jiayuguan, China. 39°48′4.64″ N, 98°12′57.84″ E. Scale bar = 100 m, ©Google Earth

Fig. 4.47 Top: Yumen Pass, China. 40°21′13.44″ N, 93°51′51.28″ E. Scale bar = 100 m, ©Google Earth. Bottom: Yumen Pass. Fort on Han Dynasty Wall (*Photo* A.S. Goudie)

Fig. 4.48 Great Wall of Gorgon, Iran. 37°15′40.08″ N, 54°58′33.54″ E. Scale bar = 100 m, ©Google Earth

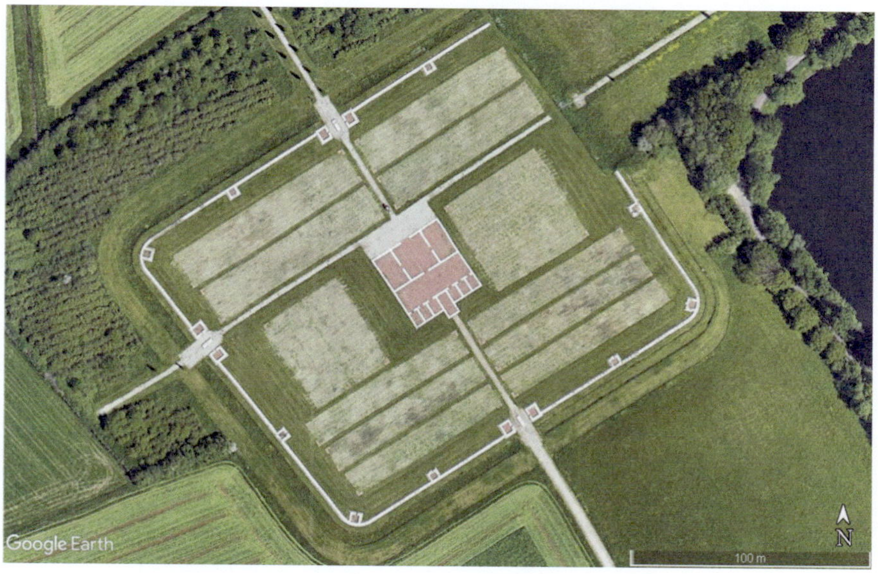

Fig. 4.49 Castellum Fectio Fort, Netherlands. 52°3′28.44″ N, 5°9′45.25″ E. Scale bar = 100 m, ©Google Earth

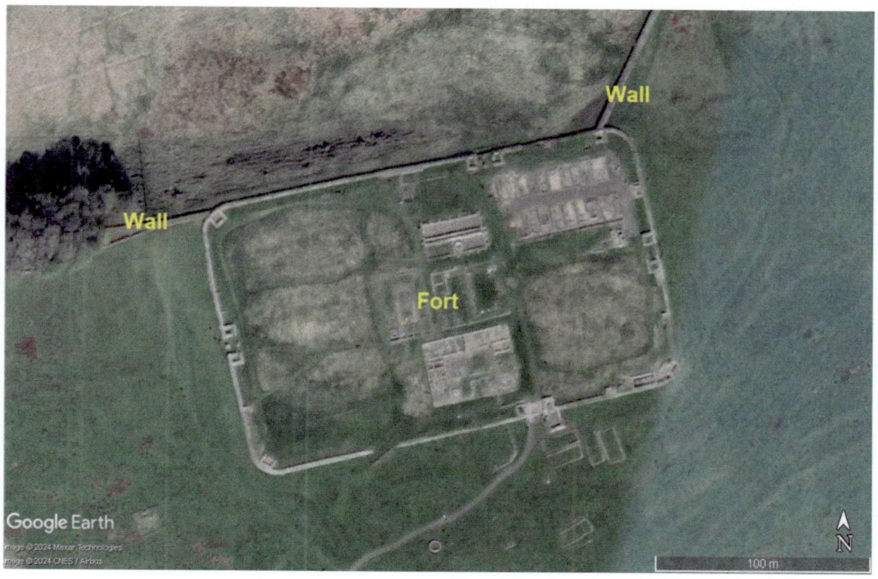

Fig. 4.50 Hadrian's Wall and Housestead Fort, England. 55°0′48.05″ N, 2°19′49.20″ W. Scale bar = 100 m, ©Google Earth

Fig. 4.51 Top: The Encircling Roman walls of Lugo, Spain. 43°0′41.81″ N, 7°33′25.77″ W. Scale bar = 300 m, ©Google Earth. Bottom: The Roman Walls of Lugo (https://commons.wikimedia.org/wiki/File:Muralla_de_Lugo_-_03.jpg) (Accessed 4th April 2025)

Romani people, the mentally ill and physically disabled, political prisoners, and those perceived as sexual deviants by the Nazi regime. All prisoners worked primarily as forced labour in local armaments factories. There were 56,545 deaths at Buchenwald of the 280,000 prisoners who passed through the camp and its subcamps. It was liberated by the US Army in April 1945, and from August 1945 to March 1950, it was used by the Soviet authorities as an internment camp. Today it serves as a Holocaust memorial.

Another type of confinement for one's political enemies is a prison. Prime examples include the Cellular Jail at Port Blair in the Andaman Islands of India, and Robben Island (where the great Nelson Mandela was incarcerated) off the Cape Coast of South Africa. The former (Fig. 4.53) was built by the British and used to exile political prisoners and Independence activists. Some of these were executed and tortured. The prison complex was constructed between 1896 and 1906. Originally built with seven wings, spanning out like a starfish from a central tower, the building suffered extensive damage during an earthquake in 1941. Later, two wings were dismantled during World War II by the Japanese invaders, and after the jail was closed in 1945 and India became independent, two more wings were demolished to make way for a nearby hospital on its northern side in the 1950s. Today, only the watchtower and three wings remain. It is now a National Monument and is on the UNESCO World

Fig. 4.52 Buchenwald Concentration Camp, Germany. 51°1′21.79″ N, 11°14′57.66″ E. Scale bar = 100 m, ©Google Earth

Fig. 4.53 Cellular Jail, Andaman Islands, India. 11°40′27.53″ N, 92°44′54.04″ E. Scale bar = 100 m, ©Google Earth. This shows the three remaining wings and the formerly central watchtower

Heritage Tentative List (https://whc.unesco.org/en/tentativelists/5888/) (Accessed 4th April 2025). The name, 'cellular jail', is believed to have been derived from the small, solitary cells, which prevented any prisoner from communicating with any other.

References

Abdelmohsen AM, Ismail MF (2025) Looting and archaeological destruction of the northern heritage sites in Kharga Oasis. Int J Tourism Archaeol Hospitality 5(1):93–106

Abram DR (2022) Aerial atlas of ancient Britain. Thames & Hudson, London, p 272

Al Haddad M, Ghazaleh SA, Atiyat D, Egho S (2023) Castles in Jordan: a comparative study of the Islamic and Crusader castles. Civil Eng Architect 11(4):1856–1870

Alfimov GL, Nosyrev GV, Panin V, Arzhantseva A, Oleaga G (2013) The application of cliff degradation models for estimation of the initial height of rammed-earth walls (Por-Bajin Fortress, Southern Siberia, Russia). Archaeometry 55(5):958–973

Ali SM (2011) The forts of Pakistan. Pakistaniaat J Pakistan Stud 3(2):124–150

Alibaigi S (2019) The Gawri Wall: a possible Partho-Sasanian structure in the western foothills of the Zagros Mountains. Antiquity 93:e22

Altaweel M, Zhuang Y (2018) Water societies and technologies from the past and present. University College London. https://muse.jhu.edu/book/81912. Accessed 4 April 2025.

Anderson EB (2017) The sieges of Kerak. Medieval Warfare 7(4):17–21

Clark VA (1987) The Roman castellum of Qasr Bshir. Brit Archaeol Rep Internat Series 340:457–495

Coones P, Patten J (1986) The Penguin guide to the landscape of England and Wales. Penguin Books, Harmondsworth, p 348

Coveney N (2015) Moated sites in Medieval England: a reassessment. Doctoral dissertation, University of Leicester

Cunliffe B (1970) The Saxon culture-sequence at Portchester Castle. Antiquaries J 50(1):67–85

Cunliffe B (2001) Facing the ocean. Oxford Univ Press, Oxford, p 600
Cunliffe B (1990) Before hillforts. Oxford J Archaeol 9(3):323–336
Cunliffe B (1994) After hillforts. Oxford J Archaeol 13(1):71–84
Daragan M (2020) Abrupt appearance of mega-hillforts in the Ukrainian forest-steppe in the early Scythian time: the search for an explanatory model. Tyragetia 14(1):117–139
Darvill T (1996) Prehistoric Britain from the air. Cambridge Univ Press, Cambridge, p 283
Emery FV (1962) Moated settlements in England. Geography 47:378–388
Fitzpatrick E (2009) Native enclosed settlement and the problem of the Irish 'ring-fort.' Medieval Archaeol 53(1):271–307
Frere SS, St Joseph JKS (1983) Roman Britain from the air. Cambridge Univ Press, Cambridge, p 232
Genequand D (2005) From 'desert castle' to medieval town: Qasr al-Hayr al-Sharqi (Syria). Antiquity 79:350–361
Goldsworthy A (2022) Hadrian's Wall. Apollo, New York, p 192
Hamilton S, Manley J (2001) Hillforts, monumentality and place: a chronological and topographic review of first millennium BC hillforts of south-east England. Eur J Archaeol 4(1):7–42
Hof C (2016) The late Roman city wall of Resafa/Sergiupolis (Syria). Its evolution and functional transition from representative over protective to concealing. In: Focus on fortification. New research on fortification in the Ancient Mediterranean and the Near East. Conference 6–9 December 2012 at the Danish Institute at Athens. Oxbow Books, Oxford, pp 397–412
Hyland AD (1995) The architectural history of Cape Coast. Trans Hist Soc Ghana 16(2):163–184
Jinadasa UN (2020) Changes in the cultural landscape and their impacts on heritage management: a study of Dutch Fort at Galle, Sri Lanka. Archaeological Studies Leiden University, Leiden University Press (LUP), Leiden (https://hdl.handle.net/1887/86288). Accessed 4 April 2025
Jing A (2015) A history of the Great Wall of China. SCPG, New York, p 132
Kennedy H (2001) Crusader castles. Cambridge Univ Press, Cambridge, p 242
Kennet D (1992) The towers of Ras Al Khaimah. Tribulus 22:29–30
Kenyon JR (2018) The castles of Edward I in Wales. Medieval Warfare 8(2):26–35
Kirkman J (1974) Fort Jesus: a Portuguese fortress on the East African coast. Clarendon Press, Oxford, p 327
Kuitems M et al (2020) Radiocarbon-based approach capable of subannual precision resolves the origins of the site of Por-Bajin. Proc Nat Acad Sci 117(25):14038–14041
Lilliu G (1959) The nuraghi of Sardinia. Antiquity 33:32–38
Limbert D (1996) Irish ringforts: a review of their origins. Archaeol J 153(1):243–289
Lock G (2022) Atlas of the hillforts of Britain and Ireland. Edinburgh Univ Press, Edinburgh, p 496
Mariani GS, Brandolini F, Melis RT (2024) Mining, farming, and diplomacy. Understanding the human landscape of Bronze Age Sardinia (Italy) through geospatial analysis. Quat Environ Humans: 2(5):100025
Mudar KM (1999) How many Dvaravati kingdoms? Locational analysis of first millennium AD moated settlements in central Thailand. J Anthrop Archaeol 18(1):1–28
National Trust (2015) Corfe Castle. National Trust, Swindon, p 32
Nokandeh J et al. (2006) Linear barriers of northern Iran: the great wall of Gorgan and the wall of Tammishe. Iran 44(1):121–173
Nunn PD and 8 others (2021) A Koronivalu kei Bua: Hillforts in Bua Province (Fiji), their chronology, associations, and potential significance. J Island Coast Archaeol 16(2–4):342–370
O'Reilly D, Scott G (2015) Moated sites of the Iron Age in the Mun River Valley, Thailand: new discoveries using Google Earth. Archaeol Res Asia 3:9–18
Oswald A, Dyer C, Barber M (2014) The creation of monuments: neolithic causewayed enclosures in the British Isles. Liverpool Univ Press, Liverpool, p 200
Papworth M (2011) The search for the durotriges: dorset and the west country in the late iron age. The History Press, Cheltenham, p 192
Reeve MM, Thurlby M (2005) King John's gloriette at corfe castle. J Soc Architect Hist 64(2):168–185
Rovero L, Tonietti U (2014) The Jiayuguan fortress: structural consistency and safeguarding programs. In: Mileto C, Vegas L, Soriano LG, Cristini V (eds) Earthen architecture: past, present and future. CRC Press, p 6
Royal Commission on Historical Monuments (England) (1970) An inventory of the historical monuments in the County of Dorset, vol 2, part 3. HMSO, London and Woking, p 701
Saunders NJ, Faulkner N (2010) Fire on the desert: conflict archaeology and the Great Arab Revolt in Jordan, 1916–18. Antiquity 84:514–527
Schurz WL (1918) Acapulco and the Manila Galleon. Southwest Hist Q 22(1):18–37
Scott G, O'Reilly D (2017) A re-appraisal of the spatial distribution of single and multi-moat prehistoric sites in Northeast Thailand. Archaeol Res Asia 11:69–76
Sluiter E (1949) The fortification of Acapulco, 1615–1616. Hisp Am Hist Rev 29(1):69–80
Tallet G, Bravard J-P, Garcier R (2013) The survey project at el-Deir, Kharga Oasis: first results, new hypotheses. In: Bagnall RS, Davoli P, Hope CA (eds) The oasis papers. Proceedings of the sixth international conference of the dakhleh oasis project. Oxbow Books, Oxford, pp 349–361
Wang X, Zuixiong L, Lu Z (2010) Condition, conservation, and reinforcement of the Yumen Pass and Hecang earthen ruins near Dunhuang. In: Conservation of ancient sites on the Silk Road: proceedings of the second international conference on the conservation of grotto sites, Mogao Grottoes, Dunhuang, People's Republic of China. Getty Publications, Los Angeles, pp 351–357
Winterburn JB (2021) The late Ottoman defended landscapes along the Hijaz railway in southern Jordan. Levant 53(3):366–383
Yang H, Wang S, Wang S, Zhao P, Ai M, Hu Q (2024) Moated site object detection using time series satellite imagery and an improved deep learning model in northeast Thailand. J Archaeol Sci 171:106070
Zaloga SJ (2011) The most daring raid of World War II: D-Day Pointe-du-Hoc. Rosen Publishing, New York, p 64

Burial Sites

5

Abstract

Burial sites of various morphologies are of very wide occurrence. In Bahrain there are the Dilmun mounds. In Egypt, Sudan, and China, there are large pyramids. Tumuli, which are common in the UK and Ireland, include long barrows, round barrows, bank barrows, and passage graves.

Keywords

Barrows · Tumuli · Pyramids · Dilmun mounds

5.1 Introduction

The roots of burial as a practice may reach back into the Middle Palaeolithic and coincide with the appearance of *Homo neanderthalensis* and *Homo sapiens* in Europe, the Middle East and Africa (Pettitt 2011). Evidence of burial has been found from over 100, 000 years ago. As a result, burial grounds are found throughout the world and in time they began to be associated with particular types of structures.

5.2 Burial Mounds on Bahrain

Nowhere in the world do ancient burial mounds dominate the landscape as they do in Bahrain. Archaeological research has proved that these mounds, which seen from the air have the texture of crocodile skin, were built in the Bronze Age by the inhabitants of the ancient realm of Dilmun, which managed trade between Mesopotamia, South Arabia, and India 4000 years ago. The most heavily mounded area is in northern Bahrain (Fig. 5.1), where there are as many as 172,000 mounds (Lowe 1986). Their size varies, but the majority of them measure 4.5–9 m in diameter and are 1–2 m high. Since the 1960s their number has been constantly reduced, due to the speed of economic development on the island; new roads and housing have expanded enormously with detrimental consequences. By 2006 there were perhaps no more than 20,000 left (Højlund 2007). Archaeological evidence shows that the burial sites were originally not constructed as mounds but as cylindrical low towers. They are now a World Heritage Site (https://whc.unesco.org/en/list/1542/) (accessed 8 June 2023).

5.3 Egyptian Necropolis

In the Kharga Oasis of the Western Desert in Egypt is the Necropolis of El Bagawat (Fig. 5.2). This is a Christian cemetery, and one of the oldest in the world, which functioned from the third century until the Muslim conquest occurred in the seventh century. It is not only one of the earliest, but also one of the best preserved Christian cemeteries, though as early as 1909 Beadnell noted that many had been ransacked or defaced. There are no less than 263 funerary chapels which dot the desert landscape. Most of the buildings are made of unburnt mud bricks (El-Hafez 2018). As Beadnell (1909, p. 103) remarked, 'The buildings cover a large area, and to a certain extent laid out in streets, which give the place the appearance more of a strange deserted town than of a graveyard. Some are small tombs, noticeable for their simple beauty; others are large mausolea and sanctuaries, richly ornamented with columns, pilasters and arches'.

5.4 Pyramids

Egypt is a land of pyramids, but not all of its pyramids have the classic shape. Some may be stepped, as is the case with the Step Pyramid of Pharoah Djoser at Saqqara in Egypt (Fig. 5.3). This was the first Egyptian stone pyramid to be built and marked the inauguration of this type of building.

A. Goudie, *Landscape Heritage of the World with Google Earth*, https://doi.org/10.1007/978-3-031-95555-6_5

Fig. 5.1 Top: dilmun mounds, Bahrain. 26° 8′49.69″ N, 50°30′31.34″ E. Scale bar = 100 m, ©Google Earth. Bottom: dilmun mounds in 1918 (https://commons.wikimedia.org/wiki/File:Burial_Mounds_in_Bahrain_1918.jpg) (accessed 4 April 2025). Public domain

The six-tier, four-sided structure is the earliest massive stone building in that country. It was built in the twenty-seventh century BC during the Third Dynasty for the burial of Doser. It is the central feature of a huge mortuary complex in a great courtyard.

The classic pyramids at Giza in Egypt (Fig. 5.4) (Rigano 2014) are one of the Seven Wonders of the Ancient World and the only one still standing. Located on a limestone plateau with a commanding view over the Nile, they were built by three kings of the fourth dynasty: Khufu (also known as Cheops), Khafre (Chephren), and Menkaure (Mycerinus). The pyramid of Khufu (c 2589–2566 BC), also known as The Great Pyramid, is the largest of these and for over 4000 years it was, at nearly 150 m in height, the tallest built structure in the world. The length of each side at the base averaged 230 m. The middle pyramid was built for Khafre and this structure measures 216 m on each side and was originally 143 m high. The southernmost and last pyramid to be built was that of Menkaure. Each of its sides measured 109 m, and the structure's completed height was 66 m. These pyramids were the superstructures of royal tombs, built to house the remains of dead pharaohs. They were mainly built of locally quarried limestone, along with some Aswan granite. Individual blocks weighed on average c 2.5 tonnes. The pyramids are part of a World Heritage Site (https://whc.unesco.org/en/list/86/) (accessed 4 April 2025).

Meroë (Fig. 5.5) was a large settlement in Nubia in northern Sudan. During the latter years of the first millennium BC and the early first millennium AD, the city and the Kingdom of Kush, of which Meroë was the capital, were

Fig. 5.2 Top: El Bagawat Christian Necropolis, Kharga, Egypt. 25°29′6.43″ N, 30°33′18.62″ E. Scale bar = 80 m, ©Google Earth. Bottom: El Bagawat Necropolis (*Photo* A.S. Goudie)

Fig. 5.3 Saqqara pyramid, Egypt. 29°52′16.66″ N, 31°12′59.81″ E. Scale bar = 90 m, ©Google Earth

Fig. 5.4 Top: The Pyramid of Khafre, Giza, Egypt. 29°58′33.88″ N, 31° 7′52.65″ E. Scale bar = 100 m, ©Google Earth. In the Middle Ages the stone facing of the Pyramid was removed by plunderers except at the top. Bottom: The Great Pyramid at Giza (*Photo* A.S. Goudie)

of major importance in northeast Africa. It covered an area of roughly 0.75 × 1 km, and may, at its peak, have had a population of 20,000 to 25,000 people. It was located on the east bank of the Nile around 200 km downstream to Khartoum. It had many temples, cemeteries and pyramids, and the remnants of an iron smelting industry (Humphris et al. 2018). It was a centre of trade, involving incense, ivory, ebony, and slaves. There are more than 200 pyramids in three groups. These were cemeteries for the elite. For such a settlement to thrive, and for there to be enough tree-derived charcoal for the iron industry, the area, which is now a desert, was probably better vegetated. It is now a World Heritage Site (https://whc.unesco.org/en/list/1336/) (accessed 4 April 2025).

The Maoling Mausoleum in China (Fig. 5.6) is the burial pyramid of the great Emperor Wu of Han. During his reign he led the Han Dynasty through its greatest territorial expansion. The mausoleum is located about 40 km to the west of Xi'an and is one of the western Han Dynasty imperial tombs. Construction began in 139 BC, the second year in the Emperor's reign, and took 53 years until completion upon his death. It is a trapezoidal structure built from rammed earth with a rectangular base which measures 222 by 217 m and a height of 47 m. It is one of a group of such tombs in the area (Magli 2018).

Also in China, but of later age, are the Xixia Imperial Tombs (Fig. 5.7). These are located at the eastern foothills of the Helan Mountains, west of Yinchuan. (https://whc.unesco.org/en/tentativelists/5811/https://whc.unesco.org/en/tentativelists/5811/) (accessed 4 April 2025). They are the imperial burial site of the Western Xia Dynasty (1038–1227) and one of the largest and best preserved imperial tomb complexes in China, including nine imperial tombs and 271 subordinate ones. The site was constructed during the reign of the Western Xia Dynasty, and the mausoleum complex covers an area

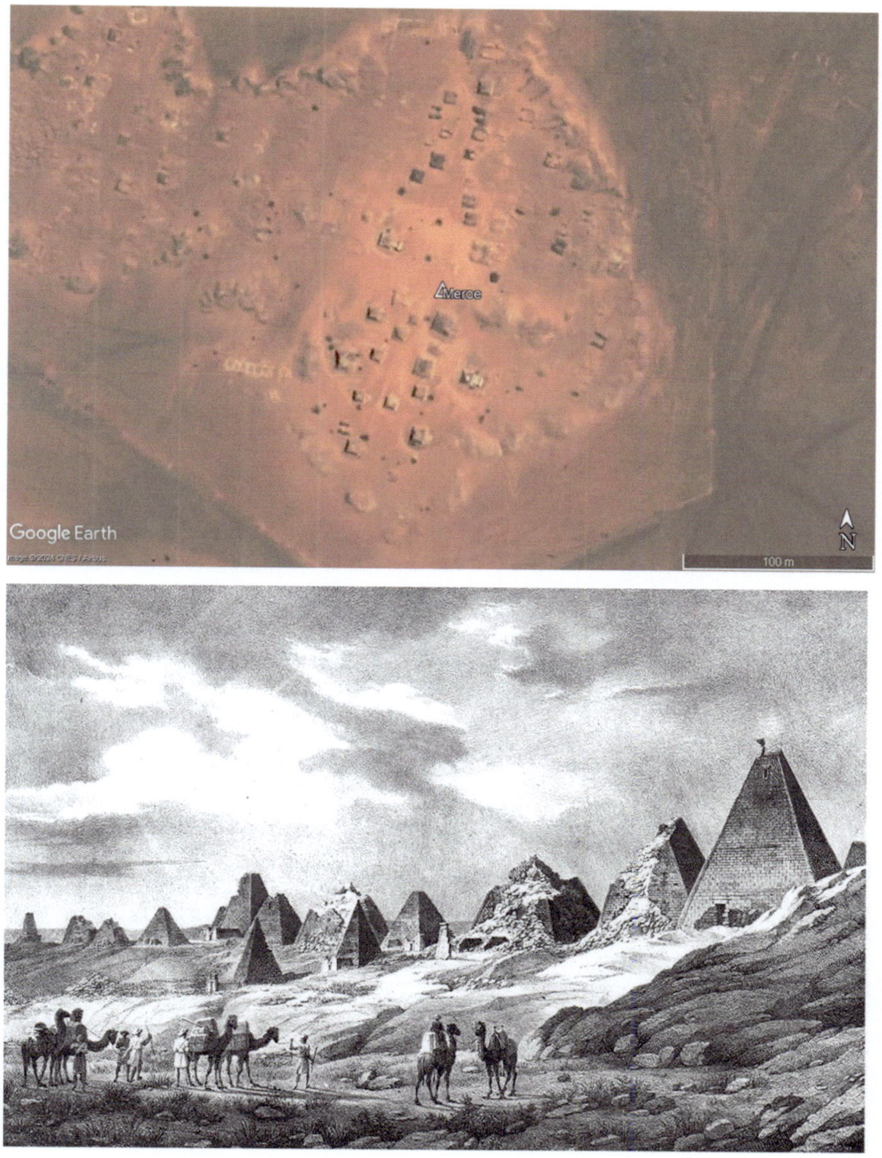

Fig. 5.5 Top: Pyramids at Meroë, Sudan. 16°55′58.03″ N, 33°43′43.77″ E. Scale bar = 100 m, ©Google Earth. Bottom: Nubian Pyramids of Meroë by Frédéric Cailliaud (1821). https://en.wikipedia.org/wiki/Pyramids_of_Mero%C3%AB#/media/File:Nubian_pyramids_of_Meroe_in_1821.jpg (accessed 4 April 2025). Public Domain. The site of Meroë was brought to the knowledge of Europeans in 1821 by Cailliaud, a French mineralogist

of approximately 50 km^2. The dynasty was founded by the Tangut people and lasted for nearly two centuries before being conquered by the Mongols in 1227. Despite wars and natural erosion, many historical relics remain intact. The mausoleums showcase a unique architectural style that blends elements of Han Chinese, Tibetan Buddhism, and Tangut culture. The imperial tombs are pyramid-like, beehive-shaped, rammed earth mound structures, earning them the nickname 'The Pyramids of the East' (Yanwu Wang, personal communication). The site includes spirit paths, stone tablets, sacrificial altars, and tomb enclosures, reflecting the distinct burial traditions of the Western Xia rulers.

In northern Peru the Moche people built impressive pyramids. Their civilization lasted from c 100 AD to 800 AD. Most notable of their creations is the Huaca del Sol (Fig. 5.8). This is an adobe brick pyramid. As time went by new layers of brick were laid directly on top of the old, hence large quantities of bricks were required for the construction—perhaps over 140 million (Hastings and Moseley 1975). It was the largest pre-Columbian adobe structure built in the Americas. It appears to have been used for a variety of ritual and ceremonial activities and to have acted as a royal residence and for burial chambers. It is believed to have originally been about 50 m high and 340 m by 160 m at its base but approximately two-thirds of the structure has been lost to erosion and looting for gold. The remaining structure we see today stands 41 m high. Archaeologists have found many remarkable remains, including face masks and portrait vessels (Bourget and Jones 2008).

Fig. 5.6 Maoling Mausoleum, China. 34°20′17.44″ N, 108°34′12.38″ E. Scale bar = 300 m, ©Google Earth

5.5 Tumuli and Barrows

Burial mounds are major landscape features in parts of Britain and Europe, where they are generally known as tumuli or barrows (https://historicengland.org.uk/images-books/publications/iha-prehistoric-barrows-burial-mounds/) (accessed 8 June 2023). An excellent discussion of these important landscape features in England and Wales is provided by Coones and Patten (1986).

Tumuli are mounds of earth and/or stone of various shapes and sizes that are characteristic prehistoric earthwork monuments dating from about 5,800 until 3,400 years ago. Less intensive and intermittent construction and use of barrow mounds also occurred in later times up until about 1,200 years ago (AD 800). Long barrows can generally be assigned to the earlier part of the time scale. Externally, they comprise a large mound of material rarely more than about 50 m in length and up to 25 m in width, but sometimes in a slightly trapezoidal or oval form, often with one end wider and higher than the other. The earliest examples developed in Iberia and western France. The tradition then spread northwards into the British Isles and then the Low Countries and southern Scandinavia. Across Europe, about 40,000 long barrows are known to survive from the Early Neolithic. They represent the oldest widespread tradition of using stone in construction. Round barrows have origins dating to before 5,000 years ago. Their size varies quite dramatically from examples of only 5 m or 6 m across to those that take on monumental proportions of over 50 m diameter and 6 m in height. The main period of round barrow construction occurred between about 4,000 and 3,500 years ago.

Tumuli associated with graves, as in Hungary, are known as *kurgans* (Dani 2020). They were built in the Ukrainian and Russian steppes, their use spreading with migration into southern, central, and northern Europe in the third millennium BC. There are also fine tumuli, the Ahom *moidams*, which date to the thirteenth to the nineteenth centuries AD, in Assam, India, and these are now a World Heritage Site (https://whc.unesco.org/en/list/1711/) (accessed 4 April 2025) (26°56′37.80″ N, 94°52′35.82″ E).

In southern England there are many areas with large number of barrows, not least in Dorset (Fig. 5.9). A huge concentration that includes round barrows, disc barrows, and bell barrows occurs in an area called Winterborne Poor Lot (Royal Commission on Historical Monuments England (1970a, pp. 461–464). These appear to have developed from 3500 BC onwards.

Nearby at the village of Long Bready there is a much rarer type of barrow—a bank barrow (Darvill 1996, pp. 160–161). The example shown in Fig. 5.10 is 196.5 m in length, but only 21 m wide. The overall height of this barrow is c 2.7 m.

Bank barrows have been described as extra-long long barrows. A more conventionally shaped feature is the long barrow at Belas Knap in Gloucestershire, England (Fig. 5.11). As seen today it is c 55 m long and 18 m wide but a good deal of reconstruction took place between 1928 and 1930. It is a chambered early Neolithic tomb of the 'Cotswold-Severn' type. It has the classic trapezoidal form and a deep forecourt at its northern end, flanked by rounded 'horns' (https://www.megalithics.com/england/belas/belamain.htm) (accessed 4 April 2025).

Widespread in Britain, Ireland, and Europe are passage graves. One example is Four Knocks Tomb in Ireland (Fig. 5.12). These were stone-built mounds covering one or more burial chambers, each accessible to the outside by means of long passages. They also occur in western Brittany in France and date to the fifth millennium BC (Cunliffe 2001, Chap. 4).

Fig. 5.7 Top: Xiaxia Imperial Tombs, China. 38°24′9.34″ N, 105°58′6.79″ E. Scale bar = 100 m, ©Google Earth. Bottom: Ground shot (*Photo* A.S.Goudie)

5.6 The Etruscans

The Etruscans occupied parts of Italy between 900 and 400 BC. They built a necropolis at Banditaccio, Cerveteri, near Rome (Fig. 5.13) (Ciuccarelli 2015). The tombs there are made by a combination of below-ground quarrying and above-ground construction with tooled blocks. The greatest belong to the seventh and sixth centuries BC and may be over 40 m in diameter. They are the burial places of leading families. In the tombs, some bodies were laid out on benches carved to resemble beds or couches with pillows at the head end. It is a World Heritage Site, and the citation states: 'The necropolis near Cerveteri, known as Banditaccia, contains thousands of tombs organized in a city-like plan, with streets, small squares and neighbourhoods. The site contains very different types of tombs: trenches cut in rock; tumuli; and some, also carved in rock, in the shape of huts or houses with a wealth of structural details. These provide the only surviving evidence of Etruscan residential architecture' (https://whc.unesco.org/en/list/1158/) (accessed 4 April 2025).

5.7 The USA, Korea, and Japan

In the Americas, there are many mounds which were constructed for a range of purposes, including for temples, burial, and settlement (Denevan 1992). In the mid-west and south of the USA, there may be as many as several hundred

Fig. 5.8 Moche Huaca del Sol, Peru. 8°7′54.19″ S, 78°59′42.19″W. Scale bar = 100 m, ©Google Earth

Fig. 5.9 Top: Winterborne Poor Lot Barrows, Dorset, England. 50°42′53.55″ N, 2°35′1.64″ W. Scale bar = 100 m, ©Google Earth. Bottom: Disc barrow (mid-ground) and round barrow (background) at Winterborne Abbas, Dorset. (*Photo* A.S. Goudie)

Fig. 5.10 Long Bready Bank Barrow, Dorset, England. 50°43′6.38″ N, 2°36′28.53″ W. Scale bar = 100 m, ©Google Earth

Fig. 5.11 Belas Knap Long Barrow, Gloucestershire, England. 51°55′38.93″ N, 1°58′15.20″ W. Scale bar = 30 m, ©Google Earth

Fig. 5.12 Four Knocks Tomb, Ireland. 53°35′47.72″ N, 6°19′35.17″ W. Scale bar = 10 m, ©Google Earth

Fig. 5.13 Banditaccia Etruscan Necropolis, Italy. 42°0′24.05″ N, 12°6′5.25″ E. Scale bar = 80 m, ©Google Earth

Fig. 5.14 Cahokia Mounds, Illinois, USA. 38°39′15.05″ N, 90°3′46.20″ W. Scale bar = 100 m, ©Google Earth

thousand artificial mounds that largely predate European colonization, and some are enormous. Notable is one at Cahokia (Fig. 5.14), near St. Louis, which is 30.5 m high and covers an area of 6.9 ha (Peet 1891). Its base is larger than the Great Pyramid at Giza in Egypt. Cahokia, a large, pre-Columbian settlement with an economy based on maize cultivation, is a World Heritage Site (https://whc.unesco.org/en/list/198/) (accessed 4 April 2025). It flourished between 1000 and 1300 AD. After that it was abandoned. The city associated with the site contained an estimated population of approximately 20,000 people with 50,000 in the surrounding area. Over 120 earthen mounds have been documented (Vilbig et al. 2020) of which 80 remain today. Today the Cahokia Mounds are considered to be the largest and most complex archaeological site north of the great pre-Columbian cities in Mexico.

In South Korea, there is a beautiful group of circular tombs at Okjeon (Fig. 5.15). They are part of a World Heritage Site (https://whc.unesco.org/en/list/1666/) (accessed 4 April 2025). The name of this cemetery, Okjeon, literally means ‘jade fields’. This is derived not only from the large amount of jade beads excavated from the cemetery, but also from the name of the neighbourhood in which it is located (Okjeon Village in Hapcheon County). Tombs housing the Daraguk ruling class were constructed here from the fourth through the mid-sixth century—Gaya Dynasty.

The largest earth-mounded tombs ever constructed were those associated with the Japanese kingdom of Yamato and were built on the plains of Osaka. The keyhole-shaped mound of Mozu Kufungun (Fig. 5.16) has three moats and is 840 m long, and occupies over 32 ha. It is associated with the Emperor Nintoku. The surface of the mound was covered

Fig. 5.15 Tumuli at Okjeon, Korea. 35°34′54.63″ N, 128°16′46.58″ E. Scale bar = 100 m, ©Google Earth

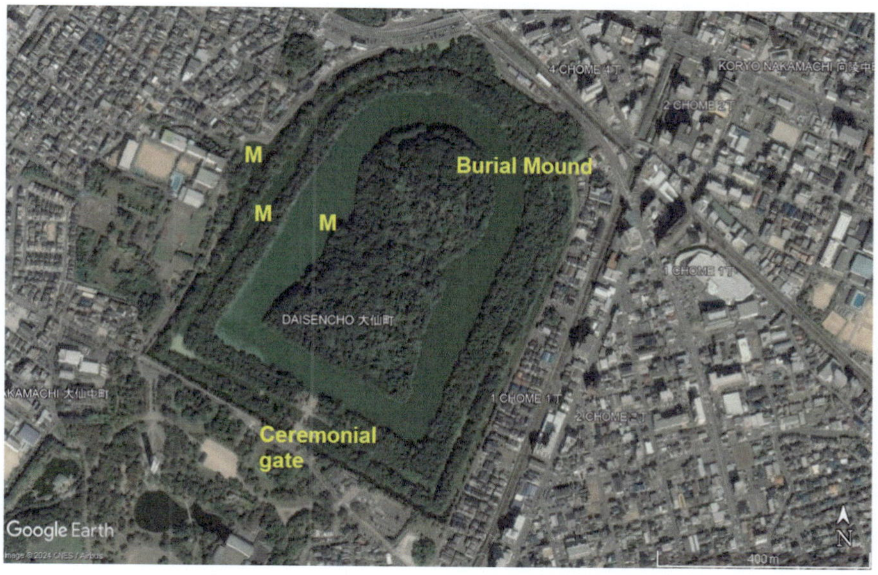

Fig. 5.16 Mozu Kufungun, Japan. 34°33′47.96″ N, 135°29′18.02″ E. Scale bar = 400 m, ©Google Earth. M = moat

with pebbles and at least 20,000 terracotta cylinders called *haniwa* (Scarre 2024, pp. 73–75). Many such mounds *(kofun)* occur in the area and were built between c 300 and 538 AD. This is a World Heritage Site (https://whc.unesco.org/en/list/1593/) (accessed 4 April 2025).

References

Beadnell HJL (1909) An Egyptian oasis. John Murray, London, p 248

Bourget S, Jones KL (eds) (2008) The art and archaeology of the Moche: an ancient Andean society of the Peruvian north coast. University of Texas Press, Austin, p 315

Ciuccarelli MR (2015) Etruscan tombs in a "Roman" city: the Necropolis of Caere between the late fourth and the first century BCE. Etruscan Studies 18(2):200–210

Coones P, Patten J (1986) The Penguin guide to the landscape of England and Wales. Penguin Books, Harmondsworth, p 348

Cunliffe B (2001) Facing the ocean. Oxford Univ Press, Oxford, p 600

Dani J (2020) Kurgans and their builders. The Great Hungarian Plain at the dawn of the Bronze Age. Hungarian Archaeol 9(2):1–20

Darvill T (1996) Prehistoric Britain from the air. Press, Cambridge, Cambridge Univ, p 283

Denevan WM (1992) The pristine myth: the landscape of the Americas in 1492. Ann Ass Am Geogr 82:369–385

El-Hafez MA (2018) Characterization and restoration recommendations of some adobe shrines at El-Bagawat Cemetery, Kharga Oasis, Western Desert-Egypt. Egyptian J Archaeol Restoration Studies 8(1):1–13

Hastings CM, Moseley ME (1975) The adobes of Huaca del Sol and Huaca de la Luna. Am Antiquity 40:196–203

Højlund F (2007) The burial mounds of Bahrain. Social complexity in Early Dilmun. Jutland Archaeological Society Publications 58

Humphris J, Charlton MF, Keen J, Sauder L, Alshishani F (2018) Iron smelting in Sudan: experimental archaeology at the royal city of Meroe. J Field Archaeol 43(5):399–416

Lowe A (1986) Bronze Age burial mounds on Bahrain. Iraq 48:73–84

Magli G (2018) Royal mausoleums of the Western Han and of the Song Chinese dynasties: a satellite imagery analysis. Archaeol Res Asia 15:45–54

Peet SD (1891) The Great Cahokia Mound. Am Antiquarian Orient J 13(1):1–31

Pettitt P (2011) The palaeolithic origins of human burial. Routledge, London, p 320

Rigano C (2014) Pyramids of the Giza Plateau. Author House, Bloomington, Indiana, p 244

Scarre C (2024) The seventy wonders of the ancient world. Thames and Hudson, London, p 336

Vilbig JM, Sagan V, Bodine C (2020) Archaeological surveying with airborne LiDAR and UAV photogrammetry: a comparative analysis at Cahokia Mounds. J Archaeol Sci: Rep 33:102509

6 Mines, Quarries, and Salterns

Abstract

In past times humans undertook quarrying for building stone and tool making, mining for minerals, excavation of peat for fuel, and exploitation of salt deposits. Key sites include granite quarries at Aswan in Egypt; flint quarries in England; peat excavations (as in the Norfolk Broads); quarries for the stones of Stonehenge; lead mines in northern England; salt ponds and mines in the Andes and Bulgaria; silver mines in Greece; nitrate mines in the Atacama Desert of Chile; shale mines in Scotland; and gold mines in Victoria, Australia.

Keywords

Quarries · Mines · Flint · Salt · Peat · Lead · Silver · Gold

6.1 Introduction

In past times, *inter alia*, humans undertook quarrying for building stone and tool making, mining for minerals, excavation of peat for fuel, and exploitation of salt deposits. Sometimes one can see the environmental results of such activities on GE, as with erosional badlands (*dongas*) in Eswatini that were possibly caused by deforestation for iron smelting as made evident by the presence of furnaces and slag (Williams 1980) (Fig. 6.1). The dongas have developed in late Pleistocene granite-derived colluvium. This contains both lime- and iron-rich layers which are needed for iron production. On the other hand, periods of colluvial erosion and deposition may have been caused by late Holocene climate changes (Lyons et al. 2024) rather than by iron smelting.

In some parts of the world there is evidence from the earliest times of stone working (e.g. for the manufacture of stone tools) but it is perhaps only from the Mesolithic onwards that actual quarries become very evident in the landscape (see, for example, Nyland 2025).

6.2 The Ancient Egyptians

One instance of this is the quarrying of stone by the ancient Egyptians. They made obelisks. These are massive monoliths which are square in section and which taper slightly to a pyramidal summit. Egyptian royal obelisks date back to c 1950 BC (the twelfth dynasty), but the largest and most impressive surviving examples are from the New Kingdom at c 1550–1069 BC. They were often associated with the worship of solar gods and were often placed in pairs flanking the pylon gateways of Egyptian temples. The stone of choice from which they were quarried was reddish granite from Aswan. An unfinished obelisk (Fig. 6.2) is found there which is thought to date to the reign of Tuthmosis II (1479–1425 BC). Had it been completed it would have been nearly 42 m high and weighing no less than 1150 tonnes. During quarrying, however, it cracked and so was abandoned. Finished obelisks, in spite of their size and weight, were transported to distant sites, including Karnak, which was c 220 km distant (Scarre 2024, pp. 272–276).

6.3 Flint Mines

At Grime's Graves in Norfolk (Fig. 6.3), eastern England, over an area of c 37 ha, there are some prehistoric flint mines which give a landscape that is dominated by a large number of craters (bell pits). In all there are c 430 of these, and they were dug with deer antler picks c 4600 years ago (Clarke 1963). They are the remains of shafts dug into the Chalk to reach seams of flint. The largest shafts are more than 14 m deep and 12 m in diameter at the surface. Low-roofed galleries supported by wooden pillars fanned out from the base of the

A. Goudie, *Landscape Heritage of the World with Google Earth*, https://doi.org/10.1007/978-3-031-95555-6_6

Fig. 6.1 Erosional badlands (donga) at Mashila in Eswatini. 26°42′22.96″ S. 31°25′46.04″ E. Scale bar = 100 m, © Google Earth. Bottom: *Donga* (badland erosion) at Mashila (*Photo* A. S. Goudie)

pits, following the flint seams sideways. Comparable features are known from elsewhere in Britain and Europe (Backowski 2014), most notably at Spiennes in Belgium and at Cissbury Ring and Harrow Hill on the South Downs of England (50°52′45.98″ N, 0°27′46.75″ W). The flint was employed to make stone axes, which were widely traded.

6.4 Stonehenge Bluestones

The megalithic circle of Stonehenge on Salisbury Plain in southern England was built in stages between 3000 and 1520 BC, during the transition from the Neolithic to the Bronze Age. It is composed of two main types of stone. The first of these are sarsen stones, which are derived from a site in Wiltshire c 25 km to the north, and made of a material called silcrete, which is formed in the area during the Tertiary era (Nash et al. 2020). The second type is called bluestones, and these are derived from a much greater distance. Where did its bluestones come from? Work by Pearson et al. (2019) has indicated clearly that these spotted dolerite blocks, known as 'preselite', came from the Preseli Hills (Mynydd Preseli) in north Pembrokeshire, west Wales—approximately 230 km from Salisbury Plain. One of the main quarries is called Carn Goedog (Ixer and Bevins 2017) (Fig. 6.4). How such large blocks were transported over such a long distance remains a matter of speculation.

6.5 Silver Mining in Attica

In Attica, in the vicinity of Laurion (also spelt Lavrion or Laurium), there are some ancient silver mines (Fig. 6.5). Laurion was the most important mining and metallurgical

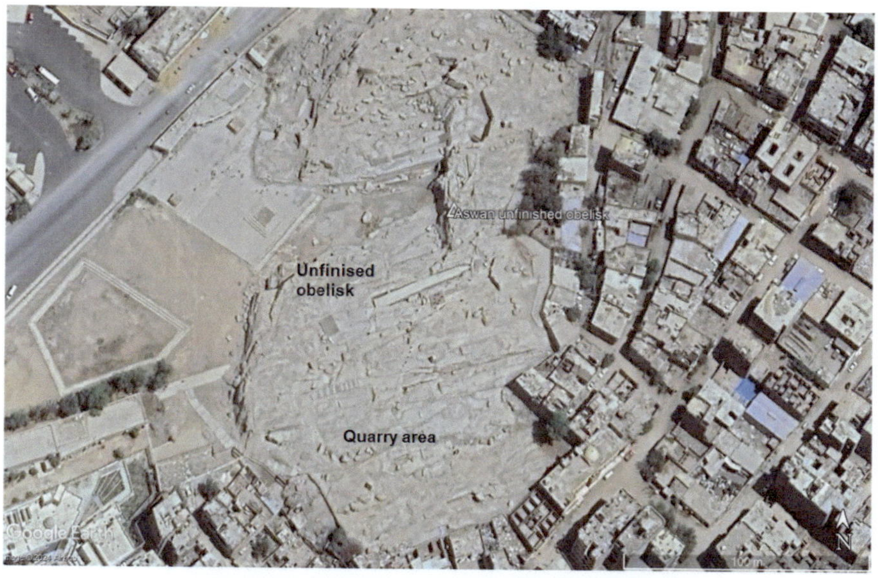

Fig. 6.2 Top: Unfinished Obelisk, Aswan, Egypt. 24°4′37.89″ N, 32°53′44.11″ E. Scale bar = 100 m, © Google Earth. Bottom: Unfinished Obelisk, Aswan (*Photo* A. S. Goudie)

centre in Ancient Greece during the fifth and fourth centuries BC. The mines produced silver, which has contributed largely to the economy of Athens and helped the Athenian city to confront the Persian invasions and to stabilize democracy. Lead was also mined. The silver from Laurion was used to fabricate many of the coins that were used throughout the region. In the fourth and fifth centuries BC, the output of the Laurion mines represented 25% of the Athenian state's annual wealth. At the beginning of the fifth century BC, the annual output was an astounding 20,000 kg of silver. Themistocles proposed that this wealth should be used to construct 200 triremes, and these were used to conduct the naval campaign against Persia which culminated in victory at the Battle of Salamis in 480 BC.

Around 700 ancient mine shafts have been discovered along with about 200 ore-processing stations. The sites show up on GE because of the presence of washeries, of which there were two types. Some were round and some were rectangular. The washeries were used to separate the dross from the ore before smelting. The crushed ore was fed into the washery where the heavier ore would sink and the lighter dross would be washed into a settling tank and the water re-used. The ore washeries were part of the workshops or *ergasteria*, along with the ore grinders, settling tanks, and water storage

Fig. 6.3 Grime's Graves, Norfolk, England. 52°28′31.67″ N, 0°40′30.46″ E. Scale bar = 100 m, © Google Earth

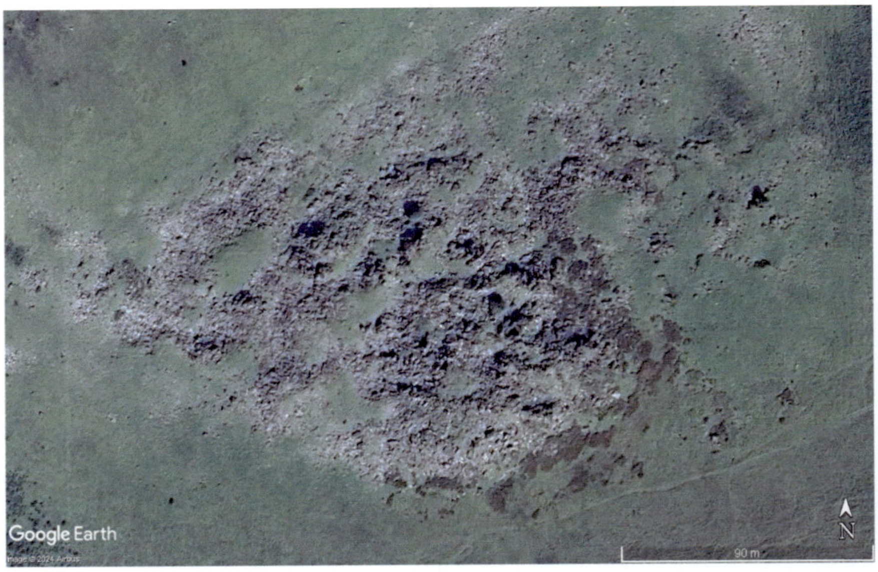

Fig. 6.4 Carn Goedog, Wales. 51°57′56.80″ N, 4°43′30.89″ W. Scale bar = 90 m, © Google Earth

tanks. A description of the washeries and their use is given by Cunningham (1967), while Jones (1982) provides detailed plans.

6.6 Lead Mining

The landscape of the Peak District in northern England reflects the fact that lead mining was an important part of the economy of the region from Roman times onwards (Kincey et al. 2022). Mine shafts, underground passages, pits, and spoil dumps occur widely (Fig. 6.6). Lead was important for such purposes as windows, typesetting, batteries, paints, water pipes, water storage, and ammunition. The height of the industry came in the seventeenth and eighteenth centuries, and the evidence is still visible today—most obviously in the form of lines of hillocks and holes (bell pits) from the more than 25,000 mine-shafts which once existed (Hose 2017). The lead ore was found in veins ('rakes') in Carboniferous Limestone (Ford and Rieuwerts 2000). The mining declined from the end of the nineteenth century in the face of cheaper and larger overseas competitors and the last remaining mines closed in the mid-twentieth century.

6.7 Diamond Mining

Particularly in southern Africa, there are the remains of former diamond mines which have now been abandoned. One of these is the Big Hole at Kimberley in South Africa (28°44′20.09″ S, 24°45′33.28″ E). Another lies in the hyper-arid Namib Desert of Namibia, near to Luderitz (Fig. 6.7). In 1908, in what was

Fig. 6.5 The Ancient Silver Mines at Laurion, Greece, showing examples of washeries. 37°41′31.08″ N, 24° 1′0.67″ E. Scale bar = 40 m, © Google Earth

Fig. 6.6 Abandoned lead mineshafts, dumps and pits, White Peak, Derbyshire, England. 53°12′33.87″ N, 1°45′41.14″ W. Scale bar = 80 m, © Google Earth. This location in c 1.25 km to the west of the famous Magpie Mine

then Deutsch-Südwestafrika, a railway worker discovered some diamonds.

Realizing the area was rich in these valuable stones, German miners soon settled, and the area was declared a *Sperrgebiet* (forbidden area). They started to exploit the diamond field. Driven by the wealth generated, they established a settlement—Kolmanskop. The residents built this in the architectural style of a German town, with a wide range of amenities and institutions including a hospital, ballroom, power station, school, skittle alley (*kegel bahn*), theatre and sports hall, casino, ice factory, railways, and tramways. However, the town started to decline during World War I as diamonds became depleted, but then was unable to compete with newly discovered diamond mines located either side of the mouth of the Orange River. The town was ultimately abandoned in 1956. A full description of the rise and fall of Kolmanskop, with contemporary maps and photographs, is given by Alexander (2010).

6.8 Salt

Salt is a necessary ingredient for life. Salt is both necessary for many aspects of physiology and crucial for cultural reasons. Salts are important dietary supplements and agents for preserving seasonal or geographically restricted meat sources, cheese and fish. Salt has been extracted in many parts of the world, either directly from ancient strata laid down in salty seas or lakes, or by evaporating water in evaporation ponds (salterns). Prehistoric salt production in Europe has

Fig. 6.7 Kolmanskop Diamond Mine, Namibia. 26°42′14.53″ S, 15°13′57.36″ E. Scale bar = 200 m, © Google Earth Top: Bottom: Abandoned personnel buildings with sand drifts (*Photo* A. S. Goudie)

been discussed in detail by Harding (2013) and in Yorkshire, England, has been detected as dating back to the Neolithic (Sherlock 2021).

The salt mines of Maras in the Andes of Peru (Fig. 6.8) are one of the largest salt extraction centres of pre-Hispanic origin in Cusco. They are located at an elevation of 3,200 m above sea level. They were developed in the Early Horizon Period (900 BC–200 BC) and were used by later groups, including the Incas. The salt has been obtained by evaporating salty water from a local subterranean stream which emerges at a spring. The flow is directed into an intricate system of small channels that were constructed so that the water runs gradually down into ponds. There is a set of approximately 4,500 salt ponds placed in the form of stepped terraces in the middle slope of a hill. The terraces are formed by retaining walls of irregular stone seated with mud mortar, forming dikes that delimit the ponds as small reservoirs of approximately 5 m across. They are a World Heritage Site (https://whc.unesco.org/en/tentativelists/6412/) (accessed 4 April 2025) and are still used for salt production.

The fortified Neolithic/Chalcolithic salt mine at Provadia in Bulgaria (Fig. 6.9) was possibly the oldest salt production centre in Europe (5500–4200 BC). The emergence and development of the 13 ha complex are directly related to its location on top of the largest deposit of rock salt in the Eastern Balkans—the so-called Mirovsko salt field or salt dome (Nikolov 2016). It is of Upper Permian age (Yanev 2000).

Salt production in salterns by using fire to evaporate seawater was very important along the British coastline in Roman and medieval times, as at Wainfleet All Saints, which is just inland of the Lincolnshire coast (Fig. 6.10) (Rippon 2000), where what are thought to be Roman saltern mounds and pits show up as intriguing patterns on GE (https://www.cambridgeairphotos.com/location/ew58/) (accessed 4 April 2025). Locally they are called 'tofts'.

One very special site is at Crosscanonbury in Cumbria. The monument includes the remains of a saltpan of post-medieval date (Fig. 6.11), and includes a large circular tank with a diameter of 17.8 m surrounded by a sea cobble and boulder wall varying in width from 0.7 to 1 m and with an entrance on its east side reinforced with sandstone blocks (https://ancientmonuments.uk/105471-allonby-saltpan-crosscanonby) (accessed 4 April 2025). Local records indicated that they were in operation from at least 1684 until c 1790 when the salt pan closed under competition from the Cheshire Triassic rock salt industry.

Fig. 6.8 Maras Salt Pans, Peru. Top: 13°17′57.90″ S, 72°9′21.41″ W. Scale bar = 100 m, © Google Earth. Bottom: https://commons.wikimedia.org/wiki/File:Salineras_de_Maras,_Maras,_Per%C3%BA,_2015-07-30,_DD_12.JPG (accessed 4 April 2025)

Fig. 6.9 Salt Mine, Provadia, Bulgaria. 43°7′42.32″ N, 27°28′22.15″ E. Scale bar = 50 m, © Google Earth

Fig. 6.10 Roman Salterns at Wainfleet All Saints, Lincolnshire, England. 53°5′33.06″ N, 0°13′51.42″ E. Scale bar = 90 m, © Google Earth

Fig. 6.11 Crosscanonbury Salt Pan, Cumbria, England. 54°44′51.09″ N, 3°27′4.06″ W. Scale bar = 20 m, © Google Earth

Many of the traditional salt pan landscapes of Mediterranean Europe have also been abandoned in recent decades in the face of industrial competition, the use of refrigeration for food preservation, and improved long-distance transport links (Hueso-Kortekaas and Carrasco-Vayá 2024; Petanidou and Dalaka 2009).

In Malta, it has been estimated that there are about 40 salinas (salt production locations) along its rocky coast. Mainland Malta has the largest percentage distribution of salinas (70%), while the remaining 30% is located in Gozo (29%) and Comino (1%). Given the massive increase in the demand for salt generated by the Knights of Malta, most typical Maltese salinas, dotting the islands' limestone coasts, date from the sixteenth to the seventeenth centuries. Most, though not all, of these are no longer active and are in highly variable states of disrepair (Gauchi et al. 2017). Traditional salinas, constructed into soft limestones on the north coast of Gozo, are shown in Fig. 6.12.

6.9 Nitrates

Just as many *salinas* have become abandoned and disused because of competition with industrial mass production, so this has happened with nitrate mines (Marr 2007). The most important of these operated in the hyper-arid Atacama Desert of Chile in the late nineteenth century. Huge quantities of saltpeter (*caliche*) were extracted at new towns, such as Humberstone (Fig. 6.13), and shipped overseas for use as fertilizers and in various other industrial processes, including the manufacture of explosives. An extensive railway network was built to take the nitrate to the ports, such as Iquique. However, in

Fig. 6.12 Xwenji Salt Pans (Salinas), Gozo, Malta. 36°4′46.27″ N, 14°14′49.42″ E. Scale bar = 30 m, © Google Earth

Fig. 6.13 Humberstone Nitrate Mine in the Atacama Desert of Chile. Top: 20°12′18.11″ S, 69°47′35.59″ W. Scale bar = 700 m, © Google Earth. Bottom: Ground Image (*Photo* A. S. Goudie)

1909, two German scientists, Fritz Haber and Carl Bosch, figured out how to chemically fix nitrogen on an industrial scale. It was disastrous for the Chilean saltpeter towns, and by 1960 they were totally abandoned. Today the town's ghostly remains stand as rusting ruins, and in 2005 they were declared a World Heritage Site (https://whc.unesco.org/en/list/1178/) (accessed 4 April 2025).

6.10 Peat Diggings

In the Norfolk landscape in eastern England, there are some freshwater lakes—the Broads (Lambert et al. 1970) (Fig. 6.14). There are 25 of these. They are of sufficient area and depth for early workers to have precluded a human origin. Later it was proposed instead that they were natural features caused by uneven alluviation and siltation of river valleys which were flooded by rapidly rising sea level in the Holocene. It was postulated by Jennings (1952) that the Broads were initiated as a series of discontinuous natural lakes. It is now clear, however, that the Broads are the result of human work (Lambert et al. 1970). Some of them have rectilinear and steep boundaries, most of them are not mentioned in early topographic books, and archival records indicate that peat-cutting (*turbary*) was widely practised in the area. On these and other grounds it is believed that peat-diggers, before AD 1300, excavated 25.5 million m^3 of peat and so created the depressions in which the lakes have formed. The flooding may have been aided by sea-level change. Comparably extensive peat excavation was also carried on in the Netherlands, notably in the fifteenth century (Smith 1966). The peat was used as a fuel.

6.11 Scottish Oil Shale Industry

A largely forgotten legacy of a former mining industry are the shale tips (*bings*) associated with the Scottish oil shale industry in West Lothian (Fig. 6.15). The Greendyke Bing has a footprint of 33 ha. Abandoned since 1925, this remarkable defunct landmark is a reminder of the oil shale industry pioneered in 1858 in West Lothian by James 'Paraffin' Young (1811–83), which soon comprised 120 oil works extracting more than 100 million litres of oil from 340 million tonnes of shale per annum, and employing up to 40,000 people. When oil production ceased there were 27 bings containing over 200 million tonnes of burnt shale (https://www.scottish-places.info/features/featurefirst18328.html) (accessed 4 April 2025). The bings are up to 95 m high. Oil shale rock was extracted from mines up to 450 m deep in Carboniferous strata. This rock was heated in retorts to extract oil vapour which was then condensed and refined to produce paraffin and other products, leaving burnt shale waste (Dean 2018; Craig and Underhill 2019).

Even more extensive are waste tips associated with former coal mining areas. There are at least 2000 million tonnes of coal waste lying in pit heaps in the coalfields of Britain, and Wales alone has c 2500 abandoned coal spoil tips (Price et al. 2011).

6.12 The Welsh Slate Industry

In North Wales, the slate industry was extremely important, particularly in the nineteenth century. Welsh slate was widely used to roof buildings in rapidly growing industrial towns and cities. This is made evident when one examines the

Fig. 6.14 The Broads, Norfolk, England, Medieval Peat Diggings. 52°42′18.79″ N, 1°27′7.64″ E. Scale bar = 700 m, © Google Earth

Fig. 6.15 Top: Greendykes Shale Bing, Scotland. 55°56′48.89″ N, 3°27′42.74″ W. Scale bar = 500 m, © Google Earth. Bottom: The nearby Broxburn Oil Shale Works (date unknown) in their heyday (https://www.britainfromabove.org.uk/en/groups) (accessed 4 April 2025)

abandoned slate quarry at Dinorwic (Fig. 6.16 top), part of a World Heritage Site (https://whc.unesco.org/en/list/1633/) (accessed 4 April 2025). Dinorwic quarry (Jones 2013) is located between Llanberis and Dinorwic. It extends over more than 283 ha of land, and at its peak, was the second largest producer of slate in the world (nearby Penrhyn was in first place) (Fig. 6.16 bottom). The first attempts to extract the Cambrian age slate at Dinorwic commenced in 1787. At its peak, in the late 1800s, the quarry employed more than 3,000 men and produced an average of 100,000 tonnes of slate per year. This was linked to the worldwide boom in demand for roofing slate roofing tiles which were exported all over the UK, Europe, and Northern America. Production of Welsh slate peaked in 1898 with 485,000 tonnes extracted. After the First World War the demand for slate had peaked and started a slow decline. By 1930 the workforce employed at the quarry had fallen to 2,000 and it continued to fall both pre- and post-Second World War. During the 1950s and 1960s it become increasingly difficult to extract any more slate from the already sheer rock galleries. Slope instability became a problem. Moreover, there was increasing competition from overseas and from cheaper tiles. The quarry therefore closed in 1969. With only 10–15% of the extracted rock providing usable slate, the quarries and mines produced millions of tonnes of waste material and these are now a striking component of the landscape (Price and Ronck 2019; Gwyn 2023), along with old tramways and the like.

6.13 The Victoria Gold Rush

In the nineteenth century gold rushes occurred in various parts of the world, including California, South Africa, Alaska, and Western Australia. One such location was Victoria in

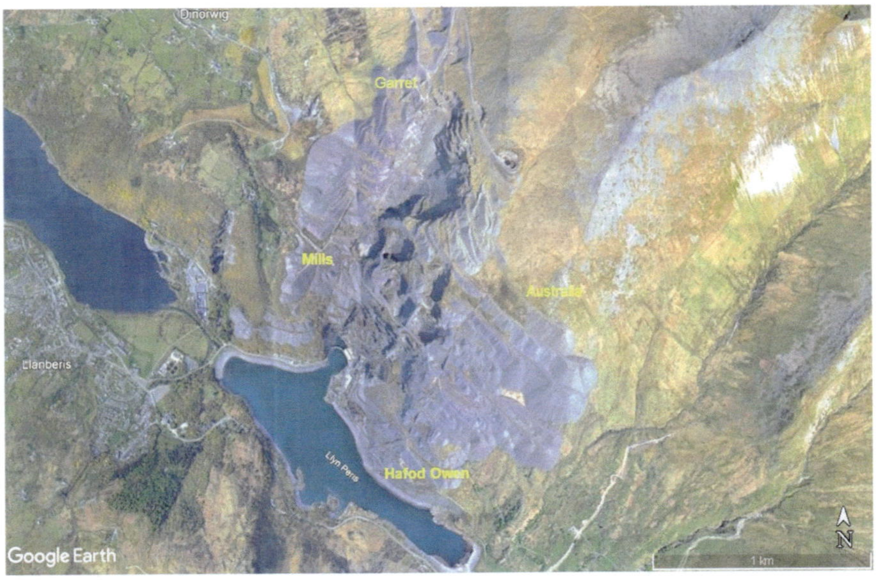

Fig. 6.16 Top: Dinorwic (Chwarel Dinorwig) Slate Quarry, Wales. 53°7′17.21″ N, 4°5′46.07″ W. Scale bar = 1 km, © Google Earth. Bottom: Penrhyn Slate Quarry in 1852 (https://en.wikipedia.org/wiki/Penrhyn_quarry#/media/File:Penrhyn_Slate_Quarries,_near_Bangor.jpeg) (accessed 4 April 2025). Public domain

Australia. Gold discovery at Ballarat in 1851 sparked that state's famous gold rush.

In Victoria there is a small, largely deserted town, called Majorca, where countless holes, marking shallow mine shafts, show the extent and intensity of gold mining that took place in fluvial gravel deposits (Fig. 6.17). Majorca is now a rural village 10 km southeast of Maryborough and 130 km northwest of Melbourne. It was founded in 1863 towards the end of the Victorian gold rush after two prospectors struck gold nearby at McCallum's Creek. Two months after the initial discovery of gold there were 250 stores and restaurants and the town had a population of about 3,000–4,000. Now the population is just a couple of hundred. The shallow gold-bearing deposits became worked out, other sources of gold employing deeper mines were developed, and so the miners moved on. Majorca was not unique. A similar landscape of holes can be found just east of McIntyre (36°41′35.06″ S, 143°42′53.04″ E).

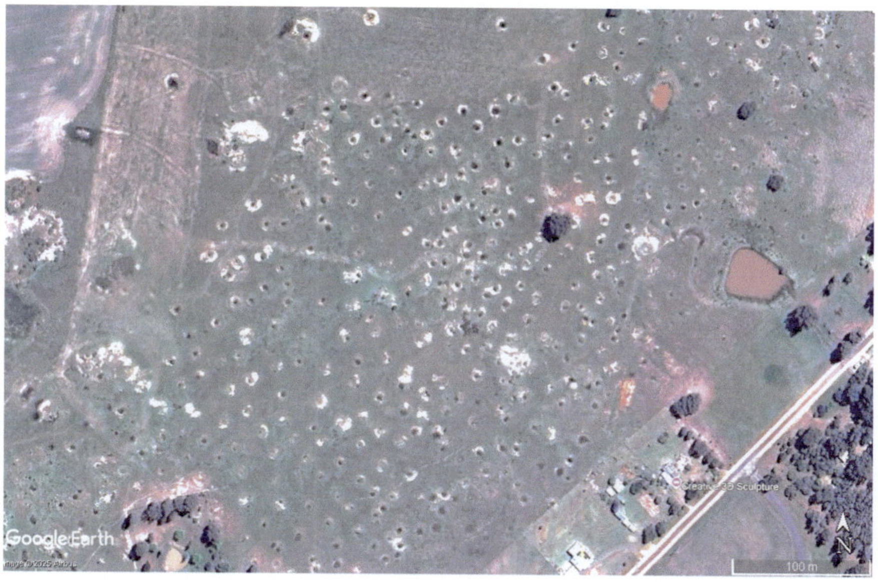

Fig. 6.17 Gold Rush Diggings, Majorca, Victoria, Australia. 37° 6′59.98″ S, 143°47′32.33″ E. Scale bar = 100 m, © Google Earth

References

Alexander N (2010) Kolmanskop: an industrial heritage resource or only a tourist attraction? Thesis, school of architecture, planning and geomatics faculty of engineering and the built environment University of Cape Town, p 125

Baczkowski J (2014) Learning by experience: The flint mines of Southern England and their continental origins. Oxford J Archaeol 33(2):135–153

Clarke RB (1963) Grime's graves, Norfolk. HMSO, London, p 32

Craig J, Underhill JR (2019) Archibald Geikie's role in the establishment of the Scottish oil shale industry. Geol Soc Lond Special Publ 480(1):379–399

Cunningham CJK (1967) The silver of Laurion. Greece & Rome 14(2):145–156

Dean G (2018) The Scottish oil-shale industry from the viewpoint of the modern-day shale-gas industry. Geol Soc Lond Special Publ 465(1):53–69

Ford TD, Rieuwerts JH (2000) Lead mining in the Peak District. Landmark Publishing, Ashbourne, p 207

Gauci R, Schembri JA, Inkpen R (2017) Traditional use of shore platforms: a study of the artisanal management of salinas on the Maltese Islands (Central Mediterranean). SAGE Open 7(2):2158244017706597

Gwyn D (2023) 'The Slate Landscape of Northwest Wales'—a new industrial World Heritage Site. Industr Archaeol Rev 45(1):19–31

Harding A (2013) Salt in prehistoric Europe. Sidestone Press, Leiden, p 162

Hose TA (2017) The English Peak District (as a potential geopark): Mining geoheritage and historical geotourism. Acta Geoturistica 8(2):32–49

Hueso-Kortekaas K, Carrasco-Vayá JF (2024) The patrimonialization of traditional salinas in Europe, a successful transformation from a productive to a services-based activity. Land 13(6):772

Ixer R, Bevins R (2017) The bluestones of Stonehenge. Geol Today 33(5):180–184

Jennings JN (1952) The origin of the Broads. Roy Geogr Soc Res Ser 2

Jones JE (1982) The Laurion silver mines: a review of recent researches and results. Greece & Rome 29(2):169–183

Jones RC (2013) Dinorwic—the Llanberis slate quarry—1780–1969. Bridge Books, Wrexham, p 176

Kincey M, Gerrard C, Warburton J (2022) Metals, mines and moorland: the changing lead mining landscapes of the North Pennines, UK, 1700–1948. Post-Medieval Archaeol 56(1):1–27

Lambert JM, Jennings JN, Smith CT, Green C, Hutchinson JN (1970) The making of the Broads: a reconsideration of their origin in the light of new evidence. Roy Geogr Soc Res Series 3:15

Lyons R, Tooth S, Duller GA, McCarthy T (2024) Are human activities or climate changes the main causes of soil erosion in the South African drylands?: a palaeo-perspective from three sites in the interior. J Quat Sci 39(7):1116–1137

Marr P (2007) Ghosts of the Atacama: the abandonment of nitrate mining in the Tarapacá region of Chile. Middle States Geogr 40:22–31

Nash DJ et al (2020) Origins of the sarsen megaliths at Stonehenge. Sci Adv 6(31):eabc0133

Nikolov V (2016) The prehistoric salt-production and urban center of Provadia-Solnitsata, northeastern Bulgaria. Méditerranée 126:71–78

Nyland AJ (2025) Quarrying and lithic procurement. In: Stutz LN, Stjerna RP, Tõrv M (eds) The Oxford handbook of Mesoithic Europe. Oxford Univ Press, Oxford, pp 798–813

Pearson MP et al (2019) Megalith quarries for Stonehenge's bluestones. Antiquity 93:45–62

Petanidou T, Dalaka A (2009) Mediterranean's changing saltscapes: a study of the abandonment of salt-making business in Greece. Global NEST J 11(4):415–433

Price WR, Ronck CL (2019) Quarrying for world heritage designation: slate tourism in North Wales. Geoheritage 4:1839–1854

Price SJ, Ford JR, Cooper AH, Neal C (2011) Humans as major geological and geomorphological agents in the Anthropocene: the significance of artificial ground in Great Britain. Phil Trans Royal Society A 369:1056–1084

Rippon S (2000) The transformation of coastal wetlands. Oxford Univ Press, Oxford, p 332

Scarre C (2024) The seventy wonders of the ancient world. Thames and Hudson, London, p 336

Sherlock SJ (2021) Early neolithic salt production at Street House, Loftus, north-east England. Antiquity 95:648–669

Smith CT (1966) Dutch peat digging and the origin of the Norfolk Broads. Geogr J 132(1):69–72

Williams DP (1980) Archaeology in Swaziland. S Afr Archaeol Bull 35(131):13–18

Yanev S (2000) Palaeozoic terranes of the Balkan Peninsula in the framework of Pangea assembly. Palaeogeogr, Palaeoclimatol, Palaeoecol 161(1–2):151–177

7 Cities and Settlements

Abstract

Urbanization has become an increasingly important phenomenon ever since the first towns and cities were founded in the Middle East. Subsequently cities were developed by the Harappans, the Egyptians, the Greeks, the Romans, the Carthaginians, the Sardinians, southern Africans (e.g. at Great Zimbabwe), the Chinese and East Asians, the great empires of Central and South America, such as those of the Aztecs, Mayans, Incas and Olmecs, and the palaeo-Indians of North America.

Keywords

Towns · Cities · Harappans · Greeks · Romans · South Americans · Palaeo-Indians

7.1 Introduction

In the early to mid-Holocene, within a few thousand years of the adoption of cereal agriculture and domestication, humans began to gather into ever larger settlements (cities and towns) and into more institutionalized social formations (states and empires). After around 6000 years ago, cities developed in the basin of the Tigris and Euphrates, and more followed by c 5000 years ago in the coastal Mediterranean, the Nile Valley, the Indus Plain, and coastal Peru. In due course, cities had evolved which had considerable human populations, and the traces of these are often revealed in today's landscapes.

7.2 Middle East

Mesopotamia is a very ancient land and is the core territory of modern Iraq. It is here that, possibly for the first time, villages became towns, and that writing, the essential tool of civilization, was invented. Various groups ruled Babylon (Fig. 7.1), which dates back to the 3rd millennium BC, but it was in the time of the Chaldean Dynasty (625–539 BC) that Babylon, then the Neobabylonian Empire, became great. Its population may have reached 200,000, making it one of the biggest cities in the world at the time. The reign of Nebuchadnezzar II marked the apogee of the dynasty, and he was a great builder, so that many of the city remains that we see today date back to his time. They are mostly made of unbaked brick. The ziggurat, a stepped temple tower, with its six or seven stories, is especially notable. The main entrance through the inner wall was the Ishtar Gate, which probably stood over 25 m high and was decorated with bulls and dragons. The city is a World Heritage Site (https://whc.unesco.org/en/list/278/) (accessed 4 April 2025). To the west of the site is a large mound. This was built by the dictator, Saddam Hussein, in the late twentieth century. The city's history is described in detail by Beaulieu (2018).

Within northern Iraq, in Upper Mesopotamia, lies the ruined city of Hatra (Fig. 7.2). It is considered the richest archaeological site from the Parthian Empire of which we have knowledge. It was a strongly fortified caravan city and capital of the small Arab Kingdom of Hatra. Located between the Roman and Parthian/Sasanian empires, it flourished in the 2nd AD, and was then destroyed and largely deserted.

A. Goudie, *Landscape Heritage of the World with Google Earth*, https://doi.org/10.1007/978-3-031-95555-6_7

Fig. 7.1 (top) Babylon, Iraq. 32°32′31.94″N, 44°25′17.01″E. Scale bar = 100 m, ©Google Earth. C = Courts. Bottom: Babylon Excavations 1932 (https://en.wikipedia.org/wiki/Babylon#/media/File:Babylon,_1932.jpg) (accessed 4 April 2025). Public domain

Fig. 7.2 Hatra, Iraq. 35°35′12.98″N, 42°43′11.84″E. Scale bar = 1 km, ©Google Earth

It played an important role in the Second Parthian War and withstood repeated attacks by the Romans. However, it fell shortly after 241 AD to the army of the Sassanid king Shapur I. The city was encircled by inner and outer walls nearly 2 km in diameter and fortified by more than 160 towers (Gawlikowski 1994). In the central area, the temples covered some 1.2 ha and were dominated by the Great Temple. Although some damage was achieved by IS, the site is in a good state of preservation even though UNESCO lists it as a World Heritage Site in danger (https://whc.unesco.org/en/soc/3341) (accessed 4 April 2025).

The rose-red city of Petra (Fig. 7.3) in Jordan, located within sandstone mountains between the Red and Dead Seas, was an important crossroads and trading hub (including the

Fig. 7.3 Top: The Entrance (Al Siq) to Petra, Jordan. 30°19′21.26″N, 35°27′14.09″E. Scale bar = 200 m, ©Google Earth. Middle: The Great Temple at Petra. 30°19′41.05″N, 35°26′39.38″E. Scale bar = 70 m, ©Google Earth. Bottom: Petra, Nabatean Tombs (*Photo* A.S. Goudie). Note the evidence for salt weathering, particularly at the base

frankincense trade) between Arabia, Egypt, and Phoenicia. It was established by the Nabateans around the sixth century BC. It is surrounded by hills, which gave some protection from invaders, and is entered through a narrow defile called the Siq. In AD 106, Emperor Trajan annexed the Nabatean kingdom. Its population may have reached 20,000. Nabateans both controlled the floods that afflicted the city and built dams, reservoirs, pipes, cisterns, and the like to store water in what is an arid area (Ortloff 2005). It has an array of rock-cut buildings, including The Treasury or Al-Khazneh, a Monastery (Deir), many tombs cut into rock, colonnaded streets, a garden area, and an imposing theatre. Some of the monuments show extensive weathering by salt and other processes (Heinrichs 2008). In AD 363, an earthquake destroyed many buildings and crippled the vital water management system. The first European to describe the ruins was the Swiss traveller J.L. Burckhardt during his travels in 1812. It is a World Heritage Site (https://whc.unesco.org/en/list/326/) (accessed 4 April 2025).

Persepolis (Fig. 7.4) lies around c 650 km to the south of modern-day Teheran in Iran. It is a World Heritage Site (https://whc.unesco.org/en/list/114/) (accessed 4 April 2025). It was founded by Darius I in 518 BC and was the capital of the Archaemenid Empire. The complex was taken by the army of Alexander the Great in 330 BC because he considered it to be the most hated of the cities of Asia. Built on a great platform or terraces 530 by 330 m (Humphris et al. 2018), the city's buildings, mainly constructed of local limestone, include three general types: military quarters, the treasury, and the reception and ceremonial halls and houses for the King. Noted structures include the Great Stairway, the Gate of All Nations, the great Apadana Palace, the Hall of a Hundred Columns, the

Fig. 7.4 Top: Persepolis, Iran. 29°56′4.39″N, 52°53′25.69″E. Scale bar = 200 m, ©Google Earth. Bottom: Gate of All Nations, Persepolis, Iran (https://en.wikipedia.org/wiki/Persepolis#/media/File:Gate_of_All_Nations,_Persepolis.jpg) (Accessed 4th April 2025). Public domain

Tripylon Hall and the Tachara (Darius I's palace), various other palaces, the Imperial Treasury, the Royal Stables, and the Chariot House.

Merv (Fig. 7.5), in Turkmenistan, was an oasis city on the Silk Route in Central Asia (Williams 2008). The site lies at the southern edge of the Karakum Desert, on a broad delta of rich alluvial land, created by the Murghab River which flows northwards from Afghanistan. Human settlements at Merv existed from the 3rd millennium BC until the eighteenth century AD. The site has a series of separate and adjacent walled cities. In the north of the image is one of these—the Erk Kala—which was founded in the sixth century BC. Its massive mud brick walls stand some 30 m in height and enclose an area of c 20 ha (Herrmann 1997). Later in its history, Merv was a major Islamic centre and it reached its apogee during the Muslim epoch and became a capital of the Arabic Caliphate at the beginning of ninth century and capital of the Great Seljuk Empire in the eleventh and twelfth centuries AD. Merv may then have been one of the world's largest cities, with a population of up to 500,000. However, it was destroyed by Genghis Khan's son, Tolui Khan, and the Mongols in AD 1221. The city and its irrigation system were rebuilt in the fifteenth century, but it never regained its former prosperity. The walls (12 km long) of the medieval city include towers, posterns, stairways, galleries, and in places, crenellation (a battlement with regular spaces in it through which the people inside the fort can shoot). Together the walled urban areas covered more than 1000 ha. Merv is now a World Heritage

Fig. 7.5 Top: Part of Merv, Turkmenistan. 37°39′44.97″N, 62°11′33.46″E. Scale bar = 900 m, ©Google Earth. Bottom: The Kyz Qala (Kala) Fortress, Merv (https://upload.wikimedia.org/wikipedia/commons/3/38/Turkmenistan_Merv_city.jpg) (accessed 4 April 2025). Public domain

Site (https://whc.unesco.org/en/list/886/) (accessed 4 April 2025).

7.3 Harappan

The Indus, or Harappan, civilization owes it names to the river basin where it was discovered and to the type site—Harappa (Fig. 7.6). The civilization dates to the Bronze Age and was at its height in 2600–1900 BC (Allchin and Allchin 1968). It extended from the Makran coast in the west to the Indo-Gangetic divide in the east. Major cities discussed here are Harappa itself, Mohenjo-Daro, and Dholavira.

The archaeological site of Harappa is located in the Punjab Province in northeastern Pakistan. It represents a large urban centre with at least four major mounds that represent the deposition and cultural remains of between at least 3300 BC and 1700 BC (Belcher and Belcher 2000). In the north of the site is an especially notable granary.

Mohenjo-Daro (Fig. 7.7), the 'Mound of the Dead', in Sindh Province, is the best preserved of the Harappan civilization cities, though it is crowned by a third century AD Buddhist stupa, which was built over the ruins of Harappan times. It is a World Heritage Site (https://whc.unesco.org/en/list/138/) (accessed 4 April 2025). Mohenjo-Daro was built in the twenty-sixth century BC and eventually had an estimated population of at least 40,000 people and perhaps as many as 100,000 (Lawler 2013). It prospered for several centuries, but by c. 1700 BC had been abandoned, the reasons for which have been much debated and include climatic deterioration (prolonged drought), flooding, and war. One of the great buildings for which it is famed is the waterproofed Great Bath, which is 12 × 9 m in size and 2.4 m deep. Since it was excavated in the 1920s by Sir John Marshall and in the 1950s by Rick Wheeler, the unbaked brick walls of the site have suffered from salt weathering associated with a rise in the water table caused by the spread of irrigation in the area (Goudie 1977). Attempts to deal with this threat over the last 50 years have largely foundered due to political and economic reasons (Meskell 2018, Chap. 3).

Dholavira (Fig. 7.8), in neighbouring India, is another Bronze Age city, and is located in Gujarat on an island in the Rann of Kutch. It is known locally as **Kotada** (which means large fort). The urban element of the site is defined by a city wall that measures 771 × 617 m. The city was built of mud brick and of sandstone. It dates from the third to mid-second millennium BC, and is the sixth largest of more than 1,000 Harappan sites discovered so far, and occupied for over 1,500 years. It is a World Heritage Site (https://whc.unesco.org/en/list/1645/) (accessed 4 April 2025). It has some reservoirs that were needed in this seasonally dry area. The town's sewage was disposed of through underground drains built with precisely laid bricks (Singh 2019). A catastrophic drought may have driven the final collapse of the settlement of Dholavira exactly at the onset of the dry Meghalayan (Late Holocene) stage (~4300–4100 years BP) (Sengupta et al. 2020).

7.4 Egyptian

The Egypt of the Pharaohs produced some very great cities. Most notable of all is the complex of Luxor and Karnak (Thebes) on the east bank of the River Nile (Strudwick and Strudwick 1999). At its peak it may have had a population of 50,000. In Luxor itself (Fig. 7.9) the Luxor Temple, dedicated to the deities Mut, Khonsu, and Amun, was largely constructed by the New Kingdom pharaohs Amenhotep III (1390–1352 BC) and Ramses II (1279–1213 BC), and is a

Fig. 7.6 Harappa, Pakistan, northwestern Sect. 30°37′44.88″N, 72°51′49.17″E. Scale bar = 100 m, ©Google Earth

Fig. 7.7 Top: Mohenjo-Daro, Pakistan. 27°19′37.44″N 68° 8′14.58″E. Scale bar = 50 m, ©Google Earth. Bottom: the Stupa at Mohenjo-Daro (*Photo* A.S. Goudie)

striking monument in the heart of the modern town. The pharaoh who made the most impressive additions to Amenhotep's Luxor Temple, however, was Ramses (Ramesses) II, the third pharaoh of the Nineteenth Dynasty. During his reign he built the first pylon, which became the entrance to Luxor Temple. Ramses also added six colossal statues of himself—two seated and four standing, at the temple's entrance. Modesty was not perhaps his strong point. An avenue, the Road of Rams (Fig. 7.10), which went in a straight line for about 2,700 m, linked the Luxor Temple and Karnak. It was lined with sphinxes. Karnak comprises a vast mix of temples, pylons, halls, chapels, and other buildings. Construction at this complex began during the reign of Senusret (reigned 1971–1926 BC) during the Middle Kingdom and continued into the Ptolemaic Kingdom (305–30 BC), although most of the extant buildings date from the New Kingdom (*between the sixteenth century BC and the eleventh century BC)*. Particularly grand is the Great Hypostyle Hall in the Precinct of Amun-Re. This has columns 10 to 21 m tall. Luxor and Karnak, together with the west bank sites of the Valley of the Kings and the Valley of the Queens was designated a UNESCO World Heritage Site in 1979 https://whc.unesco.org/en/list/87/) (accessed 4 April 2025).

In the Luxor/Karnak area is a place called Deir el-Medina (Fig. 7.11) (Lesko 1994). This is an ancient workers' village which was the home of the craftsmen and labourers who worked on the tombs in the Valley of the Kings, during the eighteenth to twentieth Dynasties of the New Kingdom (c

Fig. 7.8 Dholavira, Gujarat, India. 23°53′12.53″N, 70°12′48.99″E. Scale bar = 90 m, ©Google Earth

Fig. 7.9 Top: Luxor, Egypt. 25°41′58.50″N, 32°38′20.26″E. Scale bar = 100 m, ©Google Earth. Bottom: Luxor Temple (*Photo* A.S. Goudie)

Fig. 7.10 Ram sphinxes at Karnak, Egypt (*Photo* A.S. Goudie)

Fig. 7.11 Top: Deir el-Medina Workers' Village, Luxor, Egypt. 25°43′40.88″N, 32°36′5.29″E. ©Google Earth, Scale bar = 50 m. Bottom: Ground shot (*Photo* A.S. Goudie)

1550–1080 BC). It was founded by Amenhotep I (c.1541–1520 BC). The site is located in the southern part of the Theban necropolis on the west bank of the Nile, across the river from modern-day Luxor. At its peak, the village contained around 68 houses spread over a total area of 5,600 m2 with a narrow road running the length of the village. Its walls were made of mud brick, built on top of stone foundations. Houses consisted of four to five rooms. The village gradually became surrounded by chapels and temples and during the Ptolemaic period a sandstone temple dedicated to the goddess Hathor was built there. A small chapel was added to it in Roman times. The Copts later converted the temple into a church and a monastery.

7.5 Grecian

The Pan-Hellenic site of Delphi (Fig. 7.12) in Greece stands on a commanding position on the slopes of Mount Parnassus. It was regarded as the centre of the world (Scott 2014). According to mythology, it was here that the two eagles, sent out by Zeus from the ends of the universe to find the navel of the world, met. This was where one found the Delphic Oracle through which the God Apollo spoke. Apollo was the Olympian god of light, knowledge, and harmony. Delphi was both a religious centre and a substantial settlement. The Temple of Apollo, dating to the fourth century BC, was built on the remains of an earlier sixth-century temple. The Treasury was built at the end of the sixth century BC by the Athenians to house their offerings to Apollo. In addition there is the Stoa (or open-sided, covered porch) of the Athenians, built in the Ionic order, to house naval trophies of a war with Persia, a theatre, a stadium, and substantial walls. It is also noted for its statues, associated with its sporting activities, including, most famously of all, The Charioteer, and the statue of Atinous, Hadrian's boyfriend. It is a World Heritage Site (https://whc.unesco.org/en/list/393/) (accessed 4 April 2025).

Mycenae (Fig. 7.13) is one of the cities of the Mycenean civilization which dominated the eastern Mediterranean world in the Bronze Age from the fifteenth to the twelfth

Fig. 7.12 Top: Delphi, Greece. 38°28′56.49″N, 22°30′6.02″E. Scale bar = 100 m, ©Google Earth. Bottom: Antinous (https://en.wikipedia.org/wiki/Statue_of_Antinous_%28Delphi%29) and the Charioteer (https://commons.wikimedia.org/wiki/File:Auriga Delfi.jpg) (accessed 4 April 2025)

Fig. 7.13 Top: Mycenae, Greece. 37°43′49.81″N, 22°45′24.91″E. Scale bar = 100 m, ©Google Earth. Bottom: Nineteenth-century reconstruction of Mycenae by Jacob von Falke (1887), Hellas und Rom, eine Kulturgeschichte des Classischen Alterthums (W Spemann, Stuttgart, p 31)

centuries BC. It was an acropolis site built on a hill some 274 m above sea level, around 19 km inland from the Gulf of Argolis. Situated in the northeast corner of the Argive plain, it overlooked the whole area and commanded all easy routes to the Isthmus of Corinth. In addition to this strong defensive and strategic position, it had access to good farmland and an adequate water supply.The Palace at Mycenae was constructed on the summit of the hill and was surrounded by a wall of what is generally called ‘cyclopean masonry’. This is named after the mythical one-eyed giants, the Cyclopes, their supposed builders. Entry was through the famous Lion Gate. A series of beehive-shaped tombs, called *tholos*, were also built there. A key feature of the site is a *megaron* (palace) which is assumed to have been the seat of power of the ruler. With neighbouring Tiryns, it forms a World Heritage Site (https://whc.unesco.org/en/list/941/) (accessed 4 April 2025).

Although its origins go back to the Neolithic and the arrival of the first permanent settlers to Crete between 7000 and 6000 BC, Knossos (Fig. 7.14) became a place of glory and eminence in the second millennium BC (Whitley 2023). It was a great Bronze Age city and was made famous through the excavations by Sir Arthur Evans over three decades from 1900 onwards. It was he who gave the name Minoans to the early Cretans after the mythical King Minos. He also undertook a great deal of controversial restoration work (https://www.ashmolean.org/article/rebuilding-the-palace-of-minos-at-knossos) (accessed 4 April 2025). In its pomp, Knossos may have had a population of 14,000 to 18,000. The city’s prosperity was primarily based upon the

Fig. 7.14 Knossos, Crete. 35°17′53.04″N, 25° 9′47.14″E. Scale bar = 100 m, ©Google Earth

development of local resources such as oil, wine, and wool, but another factor was the expansion of trade, as made evident by the Minoan pottery found in mainland Greece, Egypt, Syria, Anatolia, Rhodes, the Cyclades, and Sicily. It flourished until around 1300 BC, when its palace was engulfed by flames (Macdonald 2022).

The wooded site of Olympia (Fig. 7.15) is in a valley in the Peloponnesus of Greece. It has been inhabited since prehistoric times (Barringer 2022). The first human settlements date back 4000 to 1100 BC. Settlements and necropolises from the Bronze Age have been unearthed along the banks of the Alpheios River. It is now a World Heritage Site (https://whc.unesco.org/en/list/517/) (accessed 4 April 2025). The citation states 'in the tenth century BC, Olympia became a centre for the worship of Zeus, and the Altis—the sacred sanctuary to the gods—has one of the highest concentrations of masterpieces from the ancient Greek world. In addition to temples, there are the remains of all the sports structures erected for the Olympic Games, which were held in Olympia every four years beginning in 776 BC'. The games lasted in the ancient form until the end of the fourth century AD, when the Roman emperor Theodosius I abolished pagan festivals. The famous stadium is evident on the east of the site. The temple of Zeus, located in the Altis, and built in 470–457 BC, is a model of the great Doric temples constructed elsewhere in the Peloponnese. This temple was constructed of a coarse local shell conglomerate, and the exposed surfaces were covered with a coat of white stucco. The most intact structure in Olympia, the Temple of Hera, was probably built about 600 BC. It is long and narrow, having 6 Doric columns across the ends and 16 along the sides.

Cyrene (Fig. 7.16), in the north of Libya, was the leading city of Cyrenaica from its foundation in the late seventh century BC to the late third century AD. Standing on the edge of an escarpment that overlooks the coastal plain, it was settled by Greeks from Thera (Santorini). Their leader, Battus, became the first king, and the dynasty ruled Cyrene for eight generations (until c. 440 BC). It was taken over by the Romans in 74 BC. It has many beautiful buildings, including a large forum and gymnasium, the great sanctuary of Apollo which has sites including the Temples of Artemis and Apollo which date back as early as the seventh century BC, and the large Temple of Zeus, built by the Greeks in 500–480 BC, and then *rebuilt by the Romans in the second century AD.* The site benefitted from the presence of springs to provide a water supply. For instance, the Ain Shahat or the Fountain of Apollo is one of the chief springs and was a fundamental water resource at Cyrene. This was particularly vital for the Sanctuary of Apollo. The Romans built aqueducts and cisterns (Abdrbba 2019). It is a World Heritage Site (https://whc.unesco.org/en/list/190/) (accessed 4 April 2025).

7.6 Roman

Throughout their empire the Romans built great cities. Baalbek (Fig. 7.17), located in modern Lebanon, is one of these. It is located at the foot of the southwest slope of the Anti-Lebanon Mountains, bordering the fertile plain of the Bekaa Valley at an altitude of 1150 m. The arrival of the Romans in Phoenicia in 64 BC led to a phase of construction. At Baalbek the first building work, that of the Sanctuary and Temple of Jupiter, was begun during the reign of Emperor Augustus in the late first century BC and was completed soon after AD 60 under Emperor Nero. The Sanctuary was lined by 104 granite columns and contained a temple surrounded by 50 other columns. The granite came all the way from Aswan in Egypt. Today only six columns remain standing, and their shadows can be seen in the GE image. The Temple of Bacchus may have been completed under Septimus Severus in the 190 s AD. It is richly and abundantly decorated and of impressive dimensions with its monumental gate sculpted with Bacchic

Fig. 7.15 Top: Olympia, Greece. 37°38′19.18″N, 21°37′52.95″E. Scale bar = 10 km, ©Google Earth. Bottom: Fallen Columns of the Temple of Zeus, Olympia in 1966 (*Photo* A.S. Goudie)

figures. Bacchus was the Roman god of agriculture, wine, and fertility. When Emperor Constantine converted the Roman Empire to Christianity in the early fourth century, a large basilica was constructed. Baalbek is a World Heritage Site (https://whc.unesco.org/en/list/294/) (accessed 4 April 2025).

Also on the fringes of the Roman Empire was Dougga (Thugga) in Tunisia (Fig. 7.18). It overlooks the Wadi Khaled Valley, 40 km southwest of Carthage. This site has various advantages: a steep hillside that is easily defended, Eocene limestone outcrops that provide the necessary building material, and a fertile valley that can be exploited for agricultural purposes. The Romans annexed Numidia in 46 BC, and the site flourished under Roman and Byzantine rule, the latter seeing the building of a large fort. However, before the annexation it was an important Punic capital. Its impressive ruins include temples, sanctuaries, a forum, a theatre, a race course, mausoleums, and public baths. It is the best preserved Roman small town in North Africa and for that reason has been

Fig. 7.16 Top: Cyrene, Libya. 32°49′5.62″N, 21°51′28.43″E. Scale bar = 80 m, ©Google Earth. Bottom: Entrance to the Gymnasium/Forum at Cyrene (*Photo* A.S. Goudie)

declared a World Heritage Site (https://whc.unesco.org/en/list/794/) (accessed 4 April 2025). Less regular in form that many Roman cities (e.g. Timgad) the Roman buildings were integrated into the sprawling Numidian ones, as seen in the bottom half of the GE image.

Leptis (Lepcis) Magna (Fig. 7.19), in Libya, was one of the greatest Roman cities in North Africa (Kenrick 2009; Matthews 2017). Menon (1972, p. 67) averred 'I would recommend any student of the past to save up Leptis Magna to the last. As a sight, as an experience, there is nothing to equal it: it satisfies completely'. It had, inter alia, a great harbour, striking public monuments, a market place, an aqueduct, imposing public baths, and a theatre. It was built during the reigns of Augustus and Tiberius and then expanded by Emperor Septimus Severus (193–211 AD), although it may have been founded much earlier by the Phoenicians. In its heyday, its population may have been 10,000 (McEvedy 2019, p. 182). Modelled on the basilica of Trajan's forum in Rome, the basilica of Leptis Magna was a massive building c 92 m long, 38 m wide, and 30.5 m high. It was conceived under Septimius Severus and completion was undertaken by his son Caracalla in 206 AD. The city remained prosperous into the fourth century AD, but declined thereafter. It is a World Heritage Site (https://whc.unesco.org/en/list/183/) (accessed 4 April 2025).

Volubilis in Morocco (Fig. 7.20) was a Mauritanian capital founded in the third century BC, which became an outpost of the Roman Empire after AD 40. The buildings of Volubilis,

Fig. 7.17 Top: Baalbek, Lebanon. 34° 0′25.19″N, 36°12′14.89″E. Scale bar = 90 m, ©Google Earth. Bottom: Temple of Bacchus, Baalbek (*Photo* A.S. Goudie)

constructed of limestone, include a town wall, monumental gates, a capitol, a basilica, baths, and triumphal arch of Caracalla. For reasons that are not clear, it was abandoned at the beginning of the reign of Diocletian in 285 AD.

Bottom: Triumphal Arch of Caracalla (*Photo* A.S. Goudie).

Thamugadi, Timgad (Fig. 7.21), in Algeria, located on the northern slopes of the Aurès Mountains in one of the drier parts of the country, is a city that was created in AD 100 by the Emperor Trajan (Lassus 1976). It has a clear orthogonal design and is a fine example of Roman city planning, though Menon (1972, p. 175) remarked that 'it is the most formidably dull piece of town-planning that can be conceived'. Buildings include a theatre, a forum, a market, and baths. After the Vandal invasion of AD 430, Timgad was destroyed. The Arab invasion brought about its final demise and it ceased to be inhabited after the eighth century. Timgad was inscribed as a UNESCO World Heritage Site in 1982 (https://whc.unesco.org/en/list/194/) (accessed 4 April 2025).

Not all Roman settlements were great cities such as those just described. There were many small settlements, including farms and villas. Great Witcombe Roman Villa (Fig. 7.22), one of the largest Roman houses in Britain, is located 8 km southeast of Gloucester at the base of the Cotswold escarpment. Ermin Street, the road between the Roman towns of Gloucester (Glevum) and Cirencester (Corinium), passed within a short distance of the villa to the northeast. Most of the remains are mid-third century AD. They include a house built

Fig. 7.18 Top: Dougga, Morocco. 36°25′19.71″N, 9°13′6.57″E. Scale bar = 100 m, ©Google Earth. Bottom: The Capitol Temple at Dougga (*Photo* A.S. Goudie). This dates to the second century AD

around three sides of a courtyard. The main feature is an octagonal room in the middle. This was probably a dining room. The east wing contains *inter alia* a kitchen and a latrine, while the west wing is largely composed of a bath suite (Holbrook 2003). To cope with the difficult terrain and landsliding the building was constructed on four terraces, cut into the hillside, and were heavily buttressed. The villa was abandoned at the end of Roman occupation. Recently, there have been suggestions that Great Witcombe may have been a site with religious significance and may even have been a temple (Walters and Rider 2023).

A Roman-British settlement from the first to third centuries AD is Chysauster in Cornwall, England. This consists of nine stone-walled courtyard houses and is well displayed on GE (50° 9′40.11″N, 5°32′24.90″W).

7.7 Carthage

Carthage (Fig. 7.23) in Tunisia is a World Heritage Site (https://whc.unesco.org/en/list/37/) (accessed 4 April 2025). The city was founded near Tunis by Phoenician traders from

Fig. 7.19 Top: Leptis Magna, Libya. 32°38′13.37″N, 14°17′41.98″E. Scale bar = 200 m, ©Google Earth. Middle: Leptis Magna Theatre in 1969 (*Photo* A.S. Goudie). Bottom: The Basilica at Leptis Magna in 1969 (*Photo* A.S. Goudie)

Tyre (in modern Lebanon) in the ninth century BC. By the sixth century BC, benefitting from Carthage's excellent harbours, they had conquered and controlled much of the southern Mediterranean—the Punic Empire. This brought them into conflict with other powers. The Punic Wars with Rome took place between 264 C and 146 BC, when the Romans finally triumphed. In the first century AD the Romans developed Carthage, and its population may have reached 500,000, but in AD 439 it was occupied by the Vandals, and in AD 637 it was captured by the Arabs and destroyed (Hoyos 2020).

7.8 Sardinia

The island of Sardinia in the Western Mediterranean is rich in archaeological remains, including some substantial ancient cities. One of these is Nora (Fig. 7.24) which lies in the

Fig. 7.20 Top: Volubilis, Morocco. 34° 4′25.13″N, 5°33′10.77″W. Scale bar = 200 m, ©Google Earth

Fig. 7.21 Timgad, Algeria. 35°29′7.17″N 6°28′6.48″E. Scale bar = 300 m, ©Google Earth

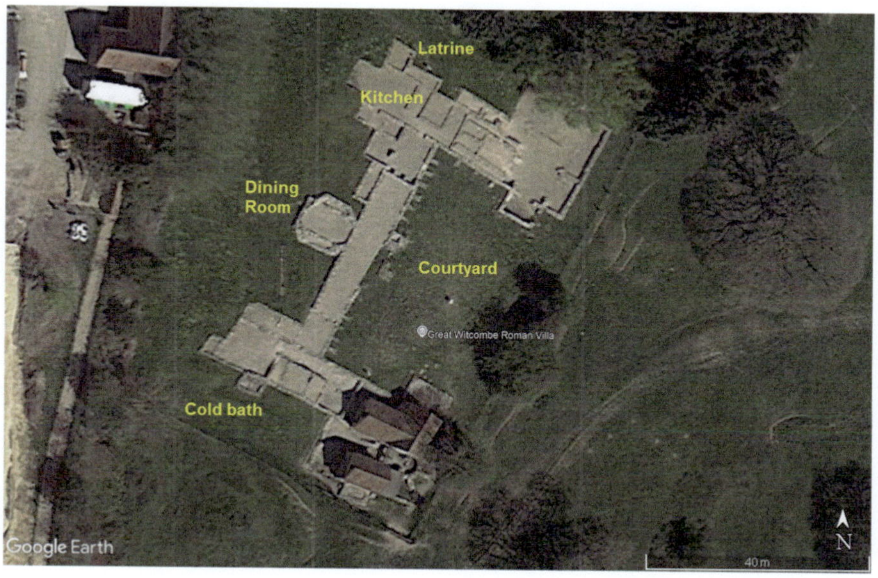

Fig. 7.22 Great Witcombe Roman Villa, Gloucestershire, England. 51°49′36.38″N, 2° 8′50.33″W. Scale bar = 40 m, ©Google Earth

far south, southwest of Cagliari. Located on a peninsula, it has two fine harbours. Its history is complex. First, there are remains of a Sardinian *nuraghe,* or towerlike monument from the Bronze Age. It then became the first Phoenician city in Sardinia (eighth century BC). It developed further in the fourth century BC under Punic rule, and was conquered by the Romans in 238 BC and became a *municipium* in the first century AD. The large theatre dates from the Roman era as do some well-preserved mosaics. Other Roman remains include an aqueduct, a temple of Juno, a nymphaeum, baths, and private villas. Under imperial Rome, Nora experienced massive urban growth and was home to around 8,000 people. Such urban development reached its peak during the third century AD. In the late fourth century AD, the construction of a Christian Basilica in the western part of the city indicated the spread of Christianity, while a steep socio-economic decline took place from the fifth until the seventh century AD. The site was probably abandoned by the end of the eighth century AD (Secco et al. 2020).

Tharros (Fig. 7.25), an ancient port city, which lies on the west coast of Sardinia, has, like Nora, a long and complex history (Acquaro and Finzi 1986). It was inhabited by Nuragic people during the period 1900–730 BC, and then was inhabited, first by Phoenicians, then by Punics, and then by Romans. Much of what one see today dates to the second and third centuries AD, when the remarkable basalt streets were laid, and the aqueduct, thermal baths, sewers, and other major monuments were built. Like Nora it has excellent mosaics (https://virtualarchaeology.sardegnacultura.it/index.php/en/archaeological-sites/eta-fenicio-punica/archaeological-area-of-tharros) (accessed 4 April 2025). Tharros was eventually abandoned around the year 1000 AD, perhaps because of pirate attacks and other raids.

7.9 Southern Africa

Great Zimbabwe (Fig. 7.26), in southern Zimbabwe, was constructed of granite, without the use of mortar, in the Iron Age by the Shona people during the eleventh to fifteenth centuries AD. It was not built by ancient Arabians and Phoenicians as some late nineteenth-century investigators had suggested. The stone-built city spans an area of 7.22 km^2 and could have housed between 10,000 and 20,000 people at its peak (c 1350—1450 AD), though estimates vary. It is said to be the largest collection of ancient ruins in Africa south of the Sahara. The Great Enclosure, with its towering free-standing walls, dates from the fourteenth century and contains a series of living quarters. Some of the walls are 11 m high. It contains a conical tower made of dry stone masonry which is 5.5 m in diameter and over 9 m high. It is not thought that these great walls were for defensive purposes but to delimit spaces, though there has been huge debate on this matter (Garlake 1973; Huffman 1984). It is a World Heritage Site (https://whc.unesco.org/en/list/364/) (accessed 4 April 2025). The reasons for the growth and then fall of Great Zimbabwe have been hotly debated (Huffman 2010; Pikirayi and Chirikure 2011). A good description of this and similar Iron Age sites is given by Mitchell (2002, Chap. 12).

Thulamela (Fig. 7.27), an Iron Age site in the far north of the Kruger Park in South Africa, covers c 9 ha, and its walls were built in the fourteenth and mid-fifteenth centuries AD (Steyn et al. 1998). It is a hilltop site overlooking the Levhuvhu River approximately 17 km from its confluence with the Limpopo River. The discovery in excavations of gold, ivory, glass, Chinese porcelain, and shell ornaments suggests that it may have had widespread trading links, even to the coast of the Indian Ocean. The city of Thulamela had a

Fig. 7.23 Top: The Military Harbour, Carthage, Tunisia. 36°50′41.65″N, 10°19′32.84″E. Scale bar = 80 m, ©Google Earth. Bottom: Carthage (*Photo* A.S. Goudie)

social system much like those of Great Zimbabwe and Mapungubwe (22°12′41.61″S, 29°23′16.53″E). About 1000 highborn people lived in a walled-in area, now called the royal enclosure or stone citadel. The surrounding hills hold traces of the homes of about 2000 commoners. Sorghum, millet, and cotton appear to have been cultivated. The site was abandoned in the early seventeenth century.

7.10 China and East Asia

Suoyang (Fig. 7.28) is a ruined city on the Silk Road in Gansu Province in northwestern China (Chen et al. 2024). It was first established in 111 BC by Emperor Wu of Han and then was relocated and rebuilt at the current site in 295 AD by Emperor Hui of the Western Jin Dynasty. It remained as an important administrative, economic, and cultural centre for over a millennium, with an estimated peak population of 50,000. It was destroyed and abandoned in the sixteenth century. Ruins of an extensive system of irrigation canals and irrigated fields remain outside of the city. They have been furrowed by *yardangs* (fluted landforms created by wind erosion) (Fan et al. 2022). The city walls, which are made of rammed earth, have been subject to frost and other weathering processes (Richards et al. 2019).

Yongtai Town (Fig. 7.29), located in Jingtai County in Gansu Province, China, was built in 1608 during the Ming Dynasty as a defensive structure. With a length of 1.7 km and

Fig. 7.24 Top: Nora, Sardinia. 38°59′3.94″N, 9° 1′0.37″E. Scale bar = 70 m, ©Google Earth. Bottom: Nora. The Roman theatre (*Photo* A.S. Goudie) (*Photo* A.S. Goudie)

a height of 12 m, the walls were made of rammed earth. It is said to have the shape of a turtle lying on the ground. It was built to defend against attacks from northern minorities and it was stationed by 2000 infantrymen and 500 cavalry units. It is now largely abandoned.

Angkor, Cambodia, is one of the finest and best-known archaeological sites in south-east Asia. Stretching over 400 km^2, the Angkor Archaeological Park contains the remains of the Khmer Empire, including the Temple of Angkor Wat (Fig. 7.30). Many of the buildings are under forest, but recently they have been mapped using LiDAR. The Empire lasted from the early ninth century AD until the fifteenth century AD. The temple was originally constructed in AD 1150 as a Hindu temple dedicated to Vishnu and was built of sandstone and laterite. A notable feature is a great moat which is 190 m wide and was gradually transformed into a Buddhist temple towards the end of that century. It has a perimeter over 5 km long. The site has many carvings of *apsaras* (female spirits) and some of the Khmer kings. It is a World Heritage Site (https://whc.unesco.org/en/list/668/) (accessed 4 April 2025). Details of the site are given in Evans and Fletcher (2015) and Fletcher et al. (2015).

7.11 South and Central America

Prior to the arrival of the Spaniards, the great empires of Central and South America, such as those of the Moche, Aztecs, Mayas, Incas, and Olmecs, built some great cities, the remains of which are well displayed on GE. Well-illustrated introductions include Jones (2020) on the Incas and Phillips (2019) on the Aztecs, Mayas, Olmecs, Mixtecs, Toltecs, and Zapotecs.

El Tajin (Fig. 7.31), located in south-east Mexico, is a city that was occupied from c 600 to 1200 AD, and became the most important settlement in Mesoamerica after the fall of the Teotihuacan Empire. The settlement had a calculated population of 15,000–20,000 inhabitants. However, El Tajín

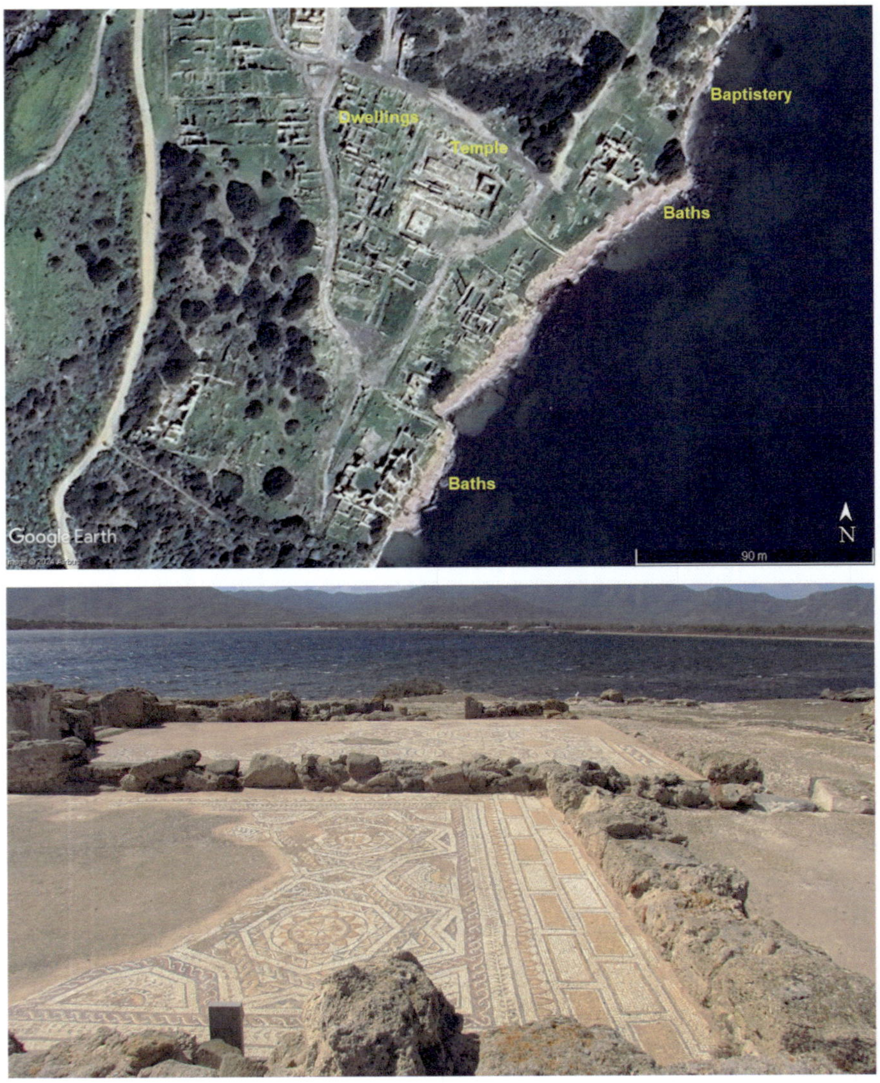

Fig. 7.25 Top: Tharros, Sardinia, Italy. 39°52′22.92″N, 8°26′27.86″E. Scale bar = 90 m, ©Google Earth. Bottom: Tharros Roman Mosaics (*Photo* A.S. Goudie)

was abandoned and partly destroyed after 1200 AD, when the region came under the rule of the powerful Aztec empire. Its Pyramid of the Niches rises in six steps to a temple at the top, with each of these having rows of square niches. There are 365 niches—one for each day of the year. Other important monuments include the Arroyo Group, the palaces of Tajín Chico, and no less than 17 ballcourts. Such courts are widespread in Mesoamerica (Scarborough and Wilcox 1993), as at Monte Albán and Chichën Itzá, but in addition to sport they were also sometimes the scene of executions. It is a World Heritage Site (https://whc.unesco.org/en/list/631/) (accessed 4 April 2025).

Monte Albán (Fig. 7.32) in the Oaxaca region of Mexico was a pre-Columbian ceremonial centre and one of the earliest and greatest cities of pre-Columbian America. It was occupied by successive civilizations—Olmecs, Zapotecs, and Mixtecs—that settled the region between 1800 BC and 900 AD. It is carved out from a mountain, and has terraces, plazas, esplanades, palaces, a court for playing ball games, *danzante* carvings, pyramids, and mounds. The city was purpose-built between 500 and 400 BC. The monumental centre of Monte Albán is the Main Plaza, which measures approximately 300 m by 150 m. Between 100 BC and 200 AD its population may have exceeded 17,000 and it was a major agricultural centre. The city also specialized in crafts, making elite produce from obsidian and seashells. The reason for its decline after c 850 AD is not clear. It is now a World Heritage Site (https://whc.unesco.org/en/list/415/) (accessed 4 April 2025).

The pre-Hispanic holy city of Teotihuacan (Fig. 7.33), named ‘the place of the Gods’ by the Aztecs, located some 48 km northeast of Mexico City, was built between the first and seventh centuries AD, though the first human establishment in the Teotihuacan area dates back to 600 BC. However, until 200 BC the site consisted of scattered small villages. From 450 to around 650 AD, it reached its peak in terms of both power and population. Around AD 700 it collapsed and was destroyed, perhaps because of belligerent neighbours

Fig. 7.26 Top: Great Zimbabwe, Zimbabwe. 20°16′20.41″S, 30°56′3.20″E. Scale bar = 100 m, ©Google Earth. Bottom: Great Enclosure, Great Zimbabwe (*Photo* A.S. Goudie)

Fig. 7.27 Thulamela, South Africa. 22°26′0.33″S, 31°11′51.71″E. Scale bar = 50 m, ©Google Earth

Fig. 7.28 Top: Suoyang, China. 40°14′38.73″N, 96°10′22.33″E. Scale bar = 1 km, ©Google Earth. Middle: Rammed earth walls at Suoyang, China (*Photo* A.S. Goudie). Bottom: Suoyang, China. 40°14′33.77″N, 96°12′52.61″E. Abandoned canals and irrigated fields, now heavily eroded by wind action. Scale bar = 300 m, ©Google Earth

Fig. 7.29 Top: Yongtai, China. 37° 8′9.83″N 103°50′49.67″E. Scale bar = 400 m, ©Google Earth. Bottom: Air photograph (*Photo*: Yanwu Wang)

Fig. 7.30 Top: Angkor Wat, Cambodia. 13°24′46.21″N, 103°52′0.65″E. Scale bar = 500 m, ©Google Earth. Bottom: Angkor Wat, Cambodia (https://commons.wikimedia.org/wiki/File:Angkor_Wat_west_gate.jpg). Public domain (accessed 4 April 2025)

Fig. 7.31 Top: El Tajin, Mexico. 20°26′48.77″N, 97°22′36.97″W. Scale bar = 200 m, ©Google Earth Bottom: The Pyramid of the Niches, El Tajin, by Carl Nebel 1836 (https://en.wikipedia.org/wiki/El_Taj%C3%ADn#/media/File:Carl_Nebel02.jpg) (accessed 4 April 2025). Public domain

Fig. 7.32 Top: Monte Albán, Mexico. 17° 2′35.14″N, 96°46′2.56″W. Scale bar = 200 m, ©Google Earth. Bottom: A danzante carving of a man, possibly a dancer, from Monte Albán. Note the huge flare earring. Olmec, Mixtec, Aztec and Zapotec men stretched their earlobes (https://commons.wikimedia.org/wiki/File:2013-12-27_Monte_Alban_Danzante_anagoria_07.JPG) (accessed 4 April 2025)

Fig. 7.33 Top: Pyramid of the Sun, Teotihuacan, Mexico. 19°41′33.69″N, 98°50′38.13″W. Scale bar = 200 m, ©Google Earth. Bottom: Teotihuacan, from photo during excavation in 1883 by D. Charney, in Charney D (1888) The ancient cities of the New World. Translated by J. Gonino and H. Conant. Harper and Brothers, New York

or because of dissatisfied insiders. Teotihuacan is the site of many of the most architecturally significant pyramids built in the pre-Columbian Americas, including the Pyramids of the Sun and the Moon (Nichols 2016). They were funerary monuments. The Temple of the Sun dominates the central city from the east side of the Avenue of the Dead. This thoroughfare is 40 m wide and 2.4 km long. The pyramid rises 66 m above-ground level, and it measures approximately c 220 by 230 m at its base. The Pyramid of the Moon is 46 m high, with a base 168 m by 149 m. The city covered 21 km^2 and at its zenith, perhaps in the first half of the first millennium, was the largest city in the Americas and one of the biggest in the world, with a population of at least 25,000 and possibly as much as 200,000. The city contained 2000 housing units, 600 pyramids, palaces, plazas, and the like. It is a World Heritage Site (https://whc.unesco.org/en/list/414/) (accessed 4 April 2025).

Chichën Itzá (Fig. 7.34) is the most important monument of the Maya-Toltec civilization in Yucatán, Mexico. It covers more than 3 km^2 and may have had a population of 30,000. The town was established around the early- or mid-fifth century, flourished between 800 and 925 AD, but rapidly declined thereafter and no major construction occurred after c 1000 AD (Ringle 2017). Major buildings include the Caracol, a circular, snail-shaped observatory (Aveni et al. 1975), and the Pyramid of Kukulcán (El Castillo). Surrounding El Castillo are terraces where various monuments were erected including The Great Ball Court (in the north west of the image), the Skull Wall, the Jaguar Temple, and the Temple of the Warriors (*Guerreros*). The buildings were connected by a dense network of paved causeways. El Castillo stands about 30 m high and consists of a series of nine square terraces, each approximately 2.6 m high. The Great Ball Court about

Fig. 7.34 Top: Chichën Itzá, Yucatan, Mexico. 20°41′1.03″N, 88°34′7.11″W. Scale bar = 100 m, ©Google Earth. Bottom: Chichën Itzá: El Castillo (https://commons.wikimedia.org/wiki/File:Chich%C3%A9n_Itz%C3%A1_pyramid.jpg) (accessed 4 April 2025)

150–200 m to the northwest of the Castillo is the largest and best preserved ball court in ancient Mesoamerica, measuring 168 by 70 m. Such ball courts, though on a less magnificent scale, are widespread in the region, as, for example, at El Tajín and Monte Albán. Water for the city was obtained from underground supplies associated with karstic development and cenote formation in the area. It is now a World Heritage Site (https://whc.unesco.org/en/list/483/) (accessed 4 April 2025). Another very significant Mayan site is Uxmal (20°21′34.04″N, 89°46′14.74″W), founded in 700 AD.

The sacred city of Caral-Supe, or simply Caral (Fig. 7.35), is a site in Peru where the remains of the main city of the Caral civilization are found. It is located in the Supe Valley near the current town of Caral, 182 km north of Lima. The Caral culture developed between 3000 and 1800 BC. It is the oldest of the pre-Hispanic civilizations (Solis et al. 2001; Solis 2006), developing 1500 years earlier than the Olmec civilization. Caral was inhabited between approximately the twenty-sixth and twentieth centuries BC, and the site covers an area of more than 60 ha. It is one of the oldest urban centres in the Americas and probably accommodated more than 3,000 inhabitants. It made extensive use of irrigation. It features complex and monumental architecture, including six large pyramidal structures. Piramide Mayor is 160 × 150 m in extent, and 18 m high. Caral was abandoned by c 1700 BC, possibly because of climatic change or earthquakes. It is now a World Heritage Site (https://whc.unesco.org/en/list/1269/) (accessed 4 April 2025).

Chan Chan (Fig. 7.36), also in Peru, was a huge pre-Columbian city some 20 km^2 in area. It was the capital of the Chimu Kingdom, which reached its apogee in the fifteenth century, not long before falling to the Incas. Chan Chan is believed to have been constructed around 850 AD and had an estimated population of 40,000–60,000 people.

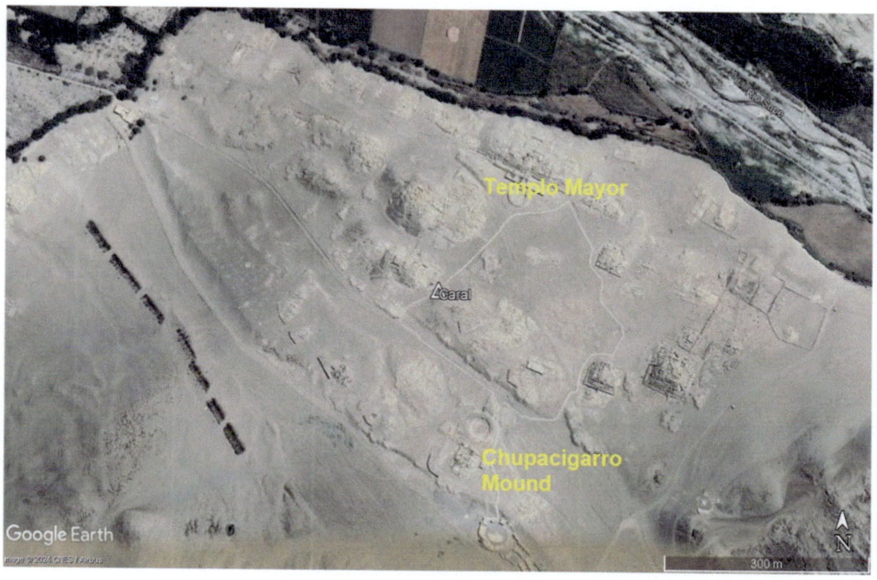

Fig. 7.35 Caral, Peru. 10°53′32.27″S, 77°31′22.68″W. Scale bar = 300 m, ©Google Earth

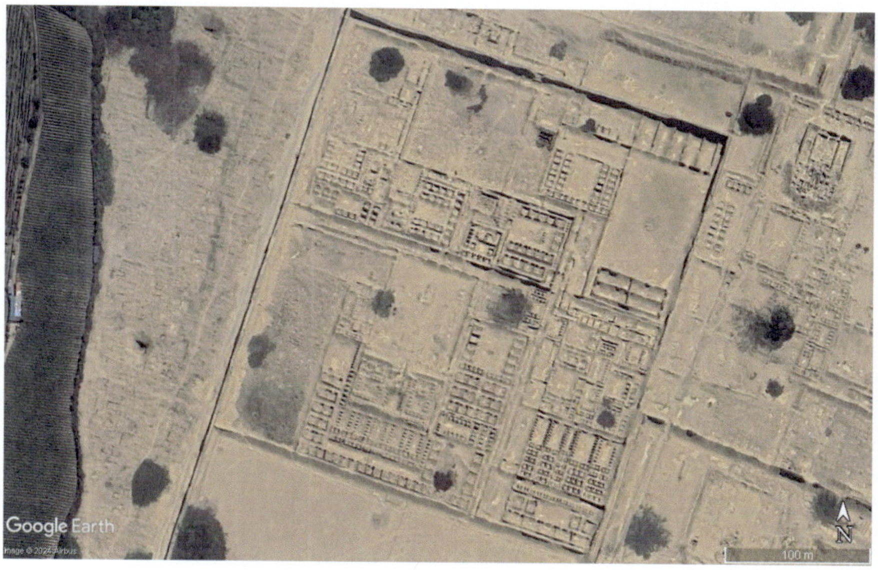

Fig. 7.36 Chan Chan, Peru. 8° 6′11.60″S, 79° 4′41.25″W. Scale bar = 100 m, ©Google Earth

It was the largest adobe city in pre-Columbian America and is now a World Heritage Site (https://whc.unesco.org/en/list/366/) (accessed 4 April 2025). It may even be the largest mud-built city in the world. It comprises a number of large rectangular complexes ('citadels' or 'palaces') delineated by high, thick earthen walls. Within these units, buildings including temples, dwellings, and storehouses are arranged around open spaces, together with reservoirs, and funeral platforms. Although the site has survived remarkably well, partly because of its location in an arid area, there are concerns that future climate changes could lead to increased rainfall erosion.

Machu Picchu (Fig. 7.37) is one of the most famous ancient city sites in South America. Located at 2430 m above sea level on a granite mountain on the eastern slopes of the Andes in Peru, its vertiginous slopes, giant dry stone walls, ramps, and terraces are remarkable. It has been called 'a royal estate in the cloud forest' (Burger and Salaazar-Burger 1993). It was built by the Incas just prior to the arrival of Europeans, and was their last stronghold. Construction appears to date from two great Inca rulers, Pachacutec Inca Yupanqui (1438–1471) and Túpac Inca Yupanqui (1472–1493). It was occupied from AD 1420 to AD 1532 (Burger et al. 2021). It is a World Heritage Site (https://whc.unesco.org/en/list/274/) (accessed

Fig. 7.37 Top: Machu Picchu, Peru. 13° 9′48.92″S, 72°32′40.63″W. Scale bar = 200 m, ©Google Earth. Middle: Machu Picchu, as photographed by Hiram Bingham in 1912 (https://en.wikipedia.org/wiki/Machu_Picchu#/media/File:Machu_Picchu.png) (accessed 4 April 2025). Public Domain. Bottom: Machu Picchu today (*Photo* Alice Goudie)

4 April 2025). Hiram Bingham, a Yale Lecturer in history, visited the site and although he was not the first to visit the ruins, it was he who brought Machu Picchu to international attention in 1911 and 1912.

7.12 North America

In pre-Columbian times, various settlements were established in North America by the palaeo-Indians. One of these, Pueblo Benito (Fig. 7.38) is associated with what is often called 'The Chaco Phenomenon', which lasted from c AD 900 to 1150 (Crown and Wills 2018). It was a time of population expansion in northwestern New Mexico. Centres based on communal masonry buildings were created and these are now known as 'great houses'. These are found throughout the San Juan Basin, but particularly in the Chaco Canyon, where several examples contained hundreds of rooms and reached four stories high. They were built of quarried sandstones, wooden timbers, and a mud mortar. The largest great house was Pueblo Bonito, which had over 600–800 rooms covering 0.8 ha to 1.2 ha. It may have had a religious and pilgrimage function for the Puebloan people, for around 37 semi-subterranean masonry-lined structures (*kivas*) have been found which were probably used primarily for ritual activities. There are also many tracks/roads in the area, some of which may have had ritual significance (Friedman et al. 2017).

The reasons for the abandonment of the site are legion and include regional climate change, deforestation, topsoil erosion, and prolonged periods of drought, building decay, and civil insurrection. These are generally cited as the main factors that lead to the abandonment of Great Houses and the extinguishing of Chacoan culture. Geologists have determined that there was indeed a 50-year-long drought that began around 1130 AD which may have ruined the agricultural base.

The site of Chetro Ketl (Fig. 7.39) lies 0.64 km from Pueblo Bonito. Like Pueblo Bonito it is an Ancestral Puebloan great house. Its construction began c 990 AD and was largely complete by 1075, but by 1250 Chetro Ketl's last inhabitants had vacated the structure. It is the second largest great house in Chaco Canyon in terms of number of rooms (approximately 400 ground and upper-storey) and the largest in surface area (more than 2.3 ha). It has some especially impressive *kivas* (ceremonial structures).

Fig. 7.38 Pueblo Bonito, New Mexico, USA. 36° 3′38.08″N, 107°57′41.36″W. Scale bar = 80 m, ©Google Earth

Fig. 7.39 Chetro Ketl, New Mexico, USA. 36° 3′37.72″N, 107°57′13.96″W. Scale bar = 50 m, ©Google Earth

References

Abdrbba MOM (2019) Water supply systems in Cyrenaica during the Greek and Roman periods: cyrene in context. Libyan Studies 50:99–105

Acquaro E, Finzi C (1986) Tharros. Carlo Defino Editore, Sassari, p 67

Allchin B, Allchin FR (1968) The birth of Indian civilization—India and Pakistan before 500 B.C. Penguin Books, Harmondsworth, p 368

Aveni AF, Gibbs SL, Hartung H (1975) The Caracol tower at Chichen Itza: an ancient astronomical observatory? Science 188:977–985

Barringer JM (2022) Olympia: a cultural history. Princeton, Princeton Univ Press, p 344

Beaulieu PA (2018) A history of Babylon, 2200 BC-AD 75. Wiley-Blackwell, Chichester, p 309

Belcher WR, Belcher WR (2000) Geologic constraints on the Harappa archaeological site, Punjab Province. Pakistan. Geoarchaeology 15(7):679–713

Burger RL, Salazar-Burger L (1993) Machu Picchu rediscovered: the royal estate in the cloud forest. Discovery 24(2):20–25

Burger RL, Salazar LC, Nesbitt J, Washburn E, Fehren-Schmitz L (2021) New AMS dates for Machu Picchu: results and implications. Antiquity 95:1265–1279

Chen W, Wu H, Guo Z, Zhang S, Qi Q, Yang L, Wu C (2024). Numerical simulation of sand dune accumulation at Suoyang Ancient City. In: Wang S, Huang R, Azzam R, Marinos VP (eds), Engineering Geology for a habitable Earth: IAEG XIV Congress 2023 Proceedings, Chengdu, China. Springer, Singapore, https://doi.org/10.1007/978-981-99-9203-4_28

Crown PL, Wills WH (2018) The complex history of Pueblo Bonito and its interpretation. Antiquity 92:890–904

Evans D, Fletcher R (2015) The landscape of Angkor Wat redefined. Antiquity 89:1402–1419

Fan Q, Liao J, Li Y, Ye W, Wang T, Yu L, ... Han W (2022) The formation of yardangs surrounding the Suoyang City ruins in the Hexi Corridor of Northwestern China and its climatic–environmental significance. Remote Sens 14(21):5628

Fletcher R, Evans D, Pottier C, Rachna C (2015) Angkor Wat: an introduction. Antiquity 89:1388–1401

Friedman RA, Sofaer A, Weiner RS (2017) Remote Sensing of Chaco roads revisited: Lidar documentation of the Great North Road, Pueblo Alto landscape, and Aztec Airport Mesa Road. Adv Archaeol Practice 5(4):365–381
Garlake P (1973) Great Zimbabwe. Thames and Hudson, London, p 224
Gawlikowski M (1994) Fortress Hatra. New evidence on ramparts and their history. Mesopotamia. Rivista di Archeologia, Epigrafia e Storia Orientale Antica 29:147–184
Goudie AS (1977) Sodium sulphate weathering and the disintegration of Mohenjo-Daro. Pakistan. Earth Surf Proc 2(1):75–86
Heinrichs K (2008) Diagnosis of weathering damage on rock-cut monuments in Petra, Jordan. Environ Geol 56:643–675
Herrmann G (1997) Early and medieval Merv: a tale of three cities. Proc Brit Acad 94:1–44
Holbrook N (2003) Great Witcombe Roman Villa, Gloucestershire. Trans Bristol Gloucestershire Archaeol Soc 121:179–200
Hoyos D (2020) Carthage: a biography. Routledge, London, p 190
Huffman TN (1984) Expressive space in the Zimbabwe culture. Man 19:593–612
Huffman TN (2010) Revisiting Great Zimbabwe. Azania 45(3):321–328
Humphris J, Charlton MF, Keen J, Sauder L, Alshishani F (2018) Iron smelting in Sudan: experimental archaeology at the royal city of Meroe. J Field Archaeol 43(5):399–416
Jones D (2020) The Inca Empire—an illustrated history. Lorenz Books, Dayton, Ohio, p 512
Kenrick P (2009) Libya archaeological guides: Tripolitania. Society for Libyan Studies, London, p 224
Lassus J (1976) Thamugadi (Timgad) Algeria. In: Stillwell R, MacDonald R, McAlister WL, Holland M (eds) The Princeton Encyclopedia of classical sites. Princeton Univ Press, Princeton. https://www.perseus.tufts.edu/hopper/text?doc=Perseus:text:1999.04.0006:entry=thamugadi. Accessed 4th April 2025
Lawler A (2013) Mohenjo-Daro's new story. Archaeology 66(1):32–37
Lesko L (ed) (1994) Pharaoh's workers: the villagers of Deir el Medina. Cornell Univ Press, Ithaca, p 224
Macdonald CF (2022) Knossos. Palatial centre of Minoan Crete. In: Norwich JJ (ed) Cities that shaped the ancient world. Thames and Hudson, London, pp 113–118
Matthews KD (2017) Cities in the sand: Leptis Magna and Sabratha in Roman Africa. Univ Pennsylvania Press, Philadelphia, p 160
McEvedy C (2019) Cities of the classical world. Penguin Books, London, p 432
Menon A (1972) Cities in the sand. Thames and Hudson, London, p 272
Meskell L (2018) A future in ruins. UNESCO, world heritage and the dream of peace. Oxford Univ Press, Oxford, p 372
Mitchell P (2002) The archaeology of southern Africa. Cambridge Univ Press, Cambridge, p 515
Nichols DL (2016) Teotihuacan. J Archaeol Res 24(1):1–74
Ortloff CR (2005) The water supply and distribution system of the Nabataean city of Petra (Jordan), 300 BC–AD 300. Cambridge Archaeol J 15(1):93–109
Phillips C (2019) Aztec and Maya—an illustrated history. Lorenz Books, Dayton, Ohio, p 512
Pikirayi I, Chirikure S (2011) Debating Great Zimbabwe. Azania 46(2):221–231
Richards J, Zhao G, Zhang H, Viles H (2019) A controlled field experiment to investigate the deterioration of earthen heritage by wind and rain. Heritage Sci 7:1–13
Ringle WM (2017) Debating Chichen Itza. Ancient Mesoam 28(1):119–136
Scarborough VL, Wilcox DR (eds) (1993) The Mesoamerican ballgame. Univ Arizona Press, Tucson, p 424
Scott M (2014) Delphi: a history of the center of the ancient world. Princeton, Princeton Univ Press, p 488
Secco M, Dilaria S, Bonetto J, Addis A, Tamburini S, Preto N, Ricci G, Artioli G (2020) Technological transfers in the Mediterranean on the verge of Romanization: Insights from the waterproofing renders of Nora (Sardinia, Italy). J Cultural Heritage 44:63–82
Singh RE (2019) Water management and conservation practices in Indus Valley civilization. Int J Applied Social Sci 6(9):2195–2199
Solis RS (2006) America's first city? The case of Late Archaic Caral. In: Isbell WH, Silverman H (eds) Andean archaeology III. Springer, Boston, MA. https://doi.org/10.1007/0-387-28940-2_3. Accessed 4th April 2025
Solis RS, Haas J, Creamer W (2001) Dating Caral, a preceramic site in the Supe Valley on the central coast of Peru. Science 292:723–726
Steyn M, Miller S, Nienaber WC, Loots M (1998) Late iron age gold burials from Thulamela (Pafuri region, Kruger National Park). South Afr Archaeol Bull 53:73–85
Strudwick N, Strudwick H (1999) Thebes in Egypt: a guide to the tombs and temples of ancient Luxor. Cornell University Press, Ithaca, New York, p 208
Walters B, Rider D (2023) Great Witcombe, Gloucestershire: a reinterpretation of the site as a temple rather than a villa. In: Henig M, Soffe G, Adcock K, King A (eds) Villas, sanctuaries and settlement in the Romano-British Countryside: New perspectives and controversies. Archaeopress, Bicester, pp 100–126
Whitley J (2023) Knossos: Myth, history and archaeology. Bloomsbury, London, p 256
Williams T (2008) The landscapes of Islamic Merv, Turkmenistan: Where to draw the line? Internet Archaeol 25:https://doi.org/10.11141/ia.25.1

8 Tells

Abstract

Long-continued settlement in roughly the same location leads over time to the accumulation of waste products, such as dung and ash, and the remains of generations of decaying mud, rammed earth, brick and stone buildings. These accumulations give rise to large mounds that are normally called *tells.* Many of the key examples come from the Middle East (e.g. Megiddo, Tell Brak, Be'er Sheva, Erbil, Kültepe, and Nisa).

Keywords

Tell · Middle East

8.1 Introduction

Long-continued settlement in roughly the same location leads over time to the accumulation of waste products, such as dung and ash, and the remains of generations of decaying mud, rammed earth, brick and stone buildings. These accumulations give rise to large mounds that are normally called *tells* (*chogha* or *tepe* in Farsi, and *hoyuk* in Turkish) (Blanco-González and Kienlin 2020). They occur in many parts of the world, including Mesoamerica (Hall 1994), West Africa (Macdonald 1997), Europe (Lubos 2011), Iran (Maghsoudi et al. 2014), India, and, above all, Mesopotamia and other long-settled parts of the Middle East (Menze and Ur 2012). Many of them date back to the beginnings of urban settlement in the early Holocene. At the present time, remote sensing has shown that many are being destroyed or looted (Elfadaly et al. 2019).

8.2 Middle Eastern Examples

Megiddo in Israel (Fig. 8.1) is an important Middle Eastern tell. It controlled a commonly used pass on the trading route between Egypt and Mesopotamia, and it also stood along the northwest-southeast route that connected the Phoenician cities with Jerusalem and the Jordan River Valley. During the Bronze Age, Megiddo was an important Canaanite city-state. During the Iron Age it was a royal city in the Kingdom of Israel, and excavations have unearthed no less than 20 strata of ruins since the Neolithic phase, indicating a very extended settlement period. The complex early stratigraphy of the site is discussed by Toffolo et al. (2014) and Martin et al. (2020).

Tell Brak in Syria (Fig. 8.2) is one of the largest ancient tells in northern Mesopotamia, over 40 m high, and 800 × 600 m in area. The main mound was occupied from at least 6000 BC to the late second millennium BC, or the end of the Late Bronze Age (Middle Assyrian Period). In its heyday, during the Late Chalcolithic period (fourth millennium BC), the site covered an area of some 110–160 ha, with a population estimate of between 17,000 and 24,000 (https://www.thoughtco.com/tell-brak-mesopotamian-capital-syria-170274) (accessed 4 April 2025).

Also in Syria is Tell Beydar (Fig. 8.3) (Wilkinson 2003, p 109–110). It appears to be of similar size to Tell Brak and is located in the far northeast of the country. They both show a long history of occupation. It was probably the site of the city of Nabada, which was first settled during the Early Dynastic Period around 2600 BC. The central site covers c 22 ha. It has a circular walled central mound (covering c 7 ha) with a circular walled lower town. Archaeologists refer to this appropriately as 'cup-and-saucer' (*kranzhugel*) morphology (Smith 2020). In Beydar there are three main phases of occupation: the first

A. Goudie, *Landscape Heritage of the World with Google Earth*, https://doi.org/10.1007/978-3-031-95555-6_8

Fig. 8.1 Megiddo Tell, Israel. 32°35′7.25″N, 35°11′4.11″E. Scale bar = 100 m, ©Google Earth

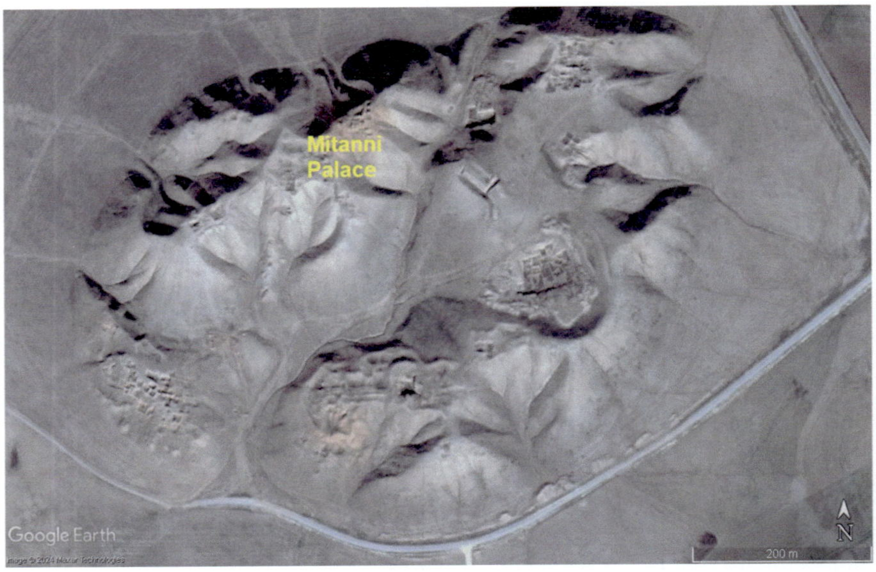

Fig. 8.2 Tell Brak, Syria. 36°40′1.77″N, 41° 3′32.23″E. Scale bar = 200 m, ©Google Earth

Fig. 8.3 Tell Beydar, Syria. 36°44′15.45″N, 40°35′15.76″E. Scale bar = 200 m, ©Google Earth

comprised the founding of the city until the decline of Nabada (c 2350 BC). The second phase witnessed the founding of a new city by the Hurrian Kingdom of Mitanni (c 1400 BC). The third phase comprised the occupation of the Assyrians in the eighth and seventh centuries BC. In recent decades, there has been considerable encroachment of buildings onto the site, as revealed by GE (Mamo et al. 2022).

In southern Israel, there is a tell at Be'er Sheva (Fig. 8.4). It rises up to 20 m above the surrounding terrain. It dates back at least to the Iron Age and there are indications of settlement during the Chalcolithic period (fourth millennium BC). It is part of a World Heritage Site (https://whc.unesco.org/en/list/1108/) (accessed 4 April 2025). It played an important role in the biblical narrative, especially the stories from the time of the Patriarchs, for it is mentioned 33 times in the Hebrew Bible. As told in Genesis, Be'er Sheva was home to Abraham, Isaac, and Jacob. It was there that Abraham formed a covenant with King Abimelech and also where both Isaac and Jacob spoke to God. However, the earliest definitive strata uncovered by archaeologists date no earlier than the Iron Age (c. 1200–1000 BC), hundreds of years after the period of the Patriarchs. It is thought that the city was probably destroyed by Sennacherib around 700 BC, and then abandoned for about 300 years. Afterwards, there is evidence of early Persian, Hellenistic, and Roman habitations, followed by an Arab presence.

Erbil (Fig. 8.5) is in the Kurdistan region of Iraq. Located on a fertile plain that supports rain-fed agriculture, Erbil and

Fig. 8.4 Beer Sheva Tell, Israel. 31°14′41.55″N, 34°50′27.62″E. Scale bar = 90 m, ©Google Earth

Fig. 8.5 Erbil, Iraq. 36°11′29.26″N, 44° 0′35.62″E. Scale bar = 200 m, ©Google Earth

its surrounds have, for thousands of years, been a regional agricultural centre, a natural gateway to the east, and a key junction on the road connecting the Arabian Gulf to the south with Anatolia to the north (Al Yaqoobi et al. 2016). In its centre is a tell and citadel. The citadel rises between 25 and 32 m from the surrounding plain. The buildings on top of the tell stretch over a roughly oval area of 430 by 340 m. It has frequently been claimed that the site is the oldest, or one of the oldest, continuously inhabited towns in the world. The earliest evidence for occupation of the citadel mound dates to the fifth millennium BC and possibly earlier. Traces of Sumerian, Assyrian, Babylonian, Greek, and Arabic settlements have all been found in the many layers beneath the present surface of the citadel. It was an important centre in the Neo-Assyrian period, especially under the reign of King Ashurbanipal 1 (669–627 BC). There are many buildings and streets of Ottoman age, and a continuous wall of tall nineteenth-century house façades conveys the visual impression of an impregnable fortress. It is a World Heritage Site (https://whc.unesco.org/en/list/1437/) (accessed 4 April 2025).

The tell at Kültepe (Fig. 8.6) (meaning 'ash hill') is located about 20 km southwest from the modern city of Kayseri in Türkiye (https://www.cappadociahistory.com/post/k%C3%BCltepe-kanesh-karum) (accessed 4 April 2025). The mound, which is 550 m long, 500 m wide, and 21 m tall, was inhabited, though with discontinuities, from the Early Bronze Age through to the Roman Empire. It was a large Assyrian trade city founded around 2000 BC and named Kanesh. This was the central city of Assyria's extensive trade network in Anatolia. Kanesh experienced great prosperity in the 1900s and 1800s BC. The hill in the middle of the mound was the imperial palace. It is on the tentative list of World Heritage Sites (https://whc.unesco.org/en/tentativelists/5905/) (accessed 4 April 2025). A full description is given by Kulakoğlu (2012).

In southern Turkmenistan, at Nisa (Fig. 8.7), there are two tells which are the remains of a city and fortress of the Parthian Empire (Lippolis 2011). A map and discussion of the site is given by Mamedov (2005). This empire was a major power from the mid-third century BC to the third century AD. The Parthians originated from the steppes south of the Aral Sea and settled in today's southern Turkmenistan, driving out the Seleucids around the middle of the third century BC. Old Nisa, the royal citadel, is a 0.14 km^2 mound surrounded by a rampart with 40 rectangular towers. New Nisa is 0.25 km^2 in area and has walls up to 9 m high. It is a World Heritage Site (https://whc.unesco.org/en/list/1242/) (accessed 4 April 2025).

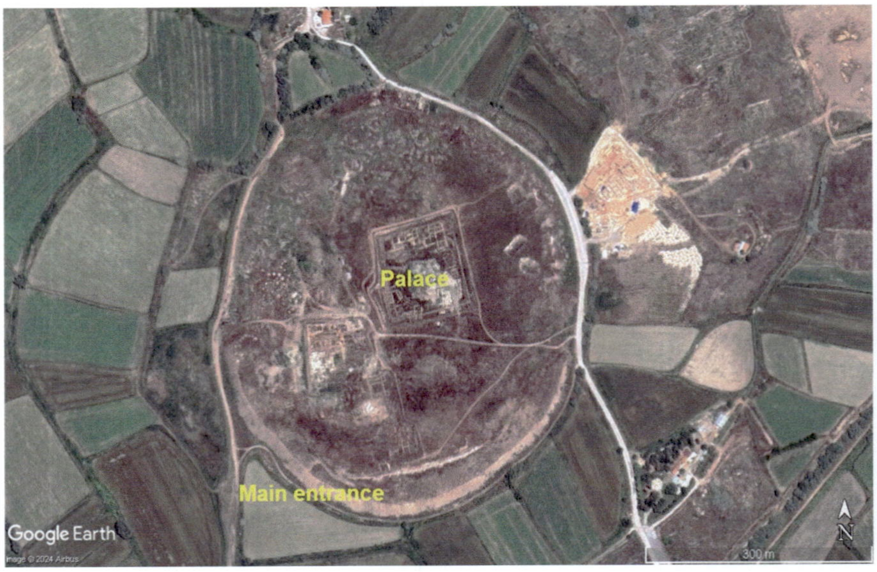

Fig. 8.6 Kültepe, Türkiye. 38°51′2.83″N, 35°38′6.75″E. Scale bar = 300 m, ©Google Earth

Fig. 8.7 Top: Old Nisa Tell, Turkmenistan. 37°57′4.45″N, 58°12′45.82″E. Scale bar = 400 m, ©Google Earth. Bottom: New Nisa Tell. 37°58′0.41″N, 58°11′58.64″E. Scale bar = 300 m, ©Google Earth

References

Al Yaqoobi D, Khorsheed A, Mohammed S, Hussein S, Shepperson M, MacGinnis J (2016) Archaeological investigations on the citadel of Erbil: background, framework and results. In: Kopanias K, MacGinnis (eds) The archaeology of the Kurdistan region of Iraq and adjacent regions. Archaeopress, Bicester, pp 23–27

Blanco-González A, Kienlin TL (eds) (2020) Current approaches to tells in the prehistoric Old World: a cross-cultural comparison from Early Neolithic to the Iron Age. Oxbow Books, Oxford, p 224

Elfadaly A, AR Abouarab M, M. El Shabrawy RR, Mostafa W, Wilson P, Morhange C, Silverstein J, Lasaponara R (2019) Discovering potential settlement areas around archaeological tells using the integration between historic topographic maps, optical, and radar data in the northern Nile Delta, Egypt. Remote Sens 11(24):3039

Hall BA (1994) Formation processes of large earthen residential mounds in La Mixtequilla, Veracruz, Mexico. Latin Am Antiquity 5:31–50

Kulakoğlu F (2012) Kültepe-Kaneš: A Second Millennium B.C.E. trading center on the Central Plateau. In: McMahon G, Steadman S (eds), The Oxford handbook of ancient Anatolia: (10,000–323 BCE). https://doi.org/10.1093/oxfordhb/9780195376142.013.0047. Accessed 14th November 2024

Lippolis C (2011) Old Nisa: the Turkmen-Italian archaeological project. Monuments of History and Culture of Turkmenistan: Discoveries Researches Restoration 20:286–301

Lubos CC, Dreibrodt S, Nelle O, Klamm M, Friederich S, Meller H, Nadeau MJ, Grootes PM, Fuchs M, Bork HR (2011) A multi-layered prehistoric settlement structure (tell?) at Niederröblingen, Germany and its implications. J Archaeol Sci 38:1101–1110

Macdonald KC (1997) More forgotten tells of Mali: an archaeologist's journey from here to Timbuktu. Archaeol Int 1:40–42

Maghsoudi M, Simpson IA, Kourampas N, Nashli HF (2014) Archaeological sediments from settlement mounds of the Sagzabad cluster, central Iran: human-induced deposition on an arid alluvial plain. Quatern Int 324:67–83

Mamedov MA, Zubanov I, Ballyev K, Moriset S, Gandreau D, Denega R, Papanov M (2005) The Parthian fortresses of Nisa, Turkmenistan: nomination of the ancient Parthian sites of Nisa for inclusion on the UNESCO World Heritage list by the Government of Turkmenistan. Doctoral dissertation, Government of Turkmenistan, p 171

Mamo AR, Ibraheem IM, Al Kassem A, Al-Khalil A, Hopper K (2022) The impact of the Syrian conflict on archaeological sites in Al-Hasakah province. J Archaeol Sci: Reports 43:103486

Martin MA, Finkelstein I, Piasetzky E (2020) Radiocarbon-dating the Late Bronze Age: cultural and historical considerations on Megiddo and beyond. Bull Am Schools Orient Res 384(1):211–240

Menze BH, Ur JA (2012) Mapping patterns of long-term settlement in northern Mesopotamia at a large scale. Proc Nat Acad Sci 109:E778-787

Smith SL (2020) The view from the steppe: Using remote sensing to Investigate the landscape of 'Kranzhügel' in its regional context. In: Lawrence D, Philip G, Altaweel M (eds) New agendas in remote sensing and landscape archaeology in the Near East: studies in honour of Tony J. Archaeopress, Oxford, Wilkinson, pp 109–123

Toffolo MB, Arie E, Martin MA, Boaretto E, Finkelstein I (2014) Absolute chronology of Megiddo, Israel, in the Late Bronze and Iron Ages: high-resolution radiocarbon dating. Radiocarbon 56(1):221–244

Wilkinson TJ (2003) Archaeological landscapes of the Middle East. Univ Arizona Press, Tucson, p 260

9 Theatres, Amphitheatres, Hippodromes, Gymnasia, and Baths

Abstract

Many classical sites are noted for their theatres and amphitheatres. Both these types of open-air venue were used for entertainment, performances, and sports. Both ancient Greek and ancient Roman theatres were built in a semicircle, with tiered seating rising on one side of the performance area. On the other hand, Roman amphitheatres were circular or oval in plan, with a central arena surrounded by perimeter seating tiers. Examples are provided from El Djem in Tunisia, Türkiye, Epidaurus in Greece, Sicily, and Libya. Sporting venues included hippodromes, gymnasia, and baths.

Keywords

Theatre · Amphitheatre · Hippodrome · Gymnasia · Baths

9.1 Introduction

Many classical sites are noted for their theatres and amphitheatres (Chase 2002). These open-air venues were used for entertainment, performances, and sports. Both ancient Greek and ancient Roman theatres were built in a semicircle, with tiered seating rising on one side of the performance area. On the other hand, Roman amphitheatres were circular or oval in plan, with a central arena surrounded by perimeter seating tiers. The seating tiers were pierced by entrance ways controlling access to the arena floor, and isolating it from the audience. The first amphitheatre may have been built at Pompeii around 70 BC. The best-known and largest Roman amphitheatre is the Colosseum in Rome itself, also known as the Flavian Amphitheatre. There was no standard size for amphitheatres. The largest could accommodate 40,000–60,000 spectators.

In addition, there were various sporting venues, which included hippodromes, gymnasia and baths, while in Chap. 7 Pre-Columbian Mesoamerican ball courts are discussed.

9.2 Amphitheatres

A spectacular and well-preserved amphitheatre is that at El Djem in Tunisia (Fig. 9.1), c 60 km south of Sousse. In classical times it was known as Thysdrus (Harris and Koser 2018). It was built by the Emperor Gordian between 230 and 238 AD, and may have accommodated in excess of 35,000 spectators. It has four floors, is elliptical in plan, and is constructed from large stone blocks. The monument's circumference is 427 m, and the full height of the structure is approximately 36 m. The ruins were declared a World Heritage Site in 1979 (https://whc.unesco.org/en/list/38/) (accessed 4 April 2025).

On mainland Italy, a large amphitheatre was that of Capua (now Santa Maria Capua Vetere) (Fig. 9.2). It was second only to the Colosseum in size and may possibly have been the model for it. With an elliptical shape, the major axis of the structure is 170 m and the minor axis is 139 m. The façade had a total height of 46 m. It was the location of the first and most famous gladiator school. It may have been built at the end of the *first century BC*.

At the other end of the Roman Empire are Maumbury Rings (Fig. 9.3), located next to Roman Dorchester (Durnovaria) in southern England. Altogether much more modest than El Djem, the earthwork, now grass-covered, is c 100 m across, and the arena area is 66 m × 49.5 m wide. This may originally have been a Neolithic henge, but was converted into an elliptical amphitheatre in the first century AD (Frere and St Joseph 1983, pp. 159–160). It would have been capable of seating up to 10,000 people in seats carved into the banks and would have been one of the biggest in England. It was modified during the Civil War in the seventeenth century to

A. Goudie, *Landscape Heritage of the World with Google Earth*, https://doi.org/10.1007/978-3-031-95555-6_9

Fig. 9.1 Top: Ej Djem Roman Amphitheatre, Tunisia. 35°17′48.11″N, 10°42′24.90″E. Scale bar = 100 m, ©Google Earth. Bottom: Amphitheatre at Ej Djem from the inside (*Photo* A.S. Goudie)

form a strong point of the town's defences (Royal Commission on Historical Monuments England 1970a, pp. 589–592). The site as it exists today is a product of remodelling during this era, with the most significant modification being a large ramp opposite the entrance. Its role as an amphitheatre was briefly revived in the late seventeenth and early eighteenth centuries, as a place for public executions. In 1685, at the close of the Monmouth Rebellion, Judge Jeffreys ordered 80 of the rebels to be executed there, and 20 years later Mary Channing, a 19-year-old woman found guilty of poisoning her husband, was executed by strangulation and burning at the Rings. It is now grassed over.

A discussion of English theatres and amphitheatres is provided by HIstoric England (2018).

9.3 Theatres

The ancient Greeks constructed theatres, with their characteristic semi-circular form. Notable among them is that at Epidaurus in the Peloponnese (Aslanidis 2003) (Fig. 9.4). It was constructed initially in the late fourth century BC, and was built in two phases: the first phase, as just noted, is dated to the late fourth century BC and the second to the mid-second century BC. It is considered to be the most perfect ancient Greek theatre with regard to both remarkable acoustics and aesthetics. The former characteristic has generated a huge literature. As it is preserved today, the theatre presents the characteristic tripartite layout of Hellenistic theatres: *cavea* (enclosure and seating area), *orchestra* and stage building.

Fig. 9.2 Capua Roman Amphitheatre, Italy. 41° 5′9.51″N, 14°15′1.58″E. Scale bar = 90 m, ©Google Earth

Fig. 9.3 Top: Maumbury Rings, Dorchester, Dorset, England. 50°42′28.85″N, 2°26′24.35″W. Scale bar = 60 m, ©Google Earth. Bottom: Maumbury Rings (*Photo:* A.S. Goudie)

Fig. 9.4 Epidaurus, Greece. 37°38′0.58″N, 23° 9′36.78″E. Scale bar = 30 m, ©Google Earth

It had a maximum capacity of 13,000 to 14,000 spectators. Because of its exceptional architecture and aesthetics, the theatre was inscribed on the World Heritage List in 1988 along with the nearby Temple of Asclepius (https://whc.unesco.org/en/list/491/) (accessed 4 April 2025).

The ancient Greeks inhabited Sicily from 750 BC, establishing many significant settlements on the island to make use of its fertile soil and advantageous trading location. Their theatre at Siracusa in southeastern Sicily (Fig. 9.5) lies on the south slopes of the Temenite Hill, overlooking the modern city. It was first built in the fifth century BC by its ruler, Hiero 1, then rebuilt in the third century BC and renovated again in the Roman period. Today, it is a part of a World Heritage Site (https://whc.unesco.org/en/list/1200/) (accessed 4 April 2025). Nearby there is a Roman amphitheatre that was built between the second half of the second and fourth centuries AD. This is said to be only surpassed in size by the Colosseum of Rome and to be even larger than the Arena di Verona.

Another important Greek theatre in Sicily is that of Taormina (Sear 1996) (Fig. 9.6). Like Siracusa, it was built in the third century BC, but under the Romans the theatre was rebuilt, probably around the time of Hadrian or Trajan. It had a diametre of 107 m and could hold around 10,000 spectators.

There are some notable theatres in Türkiye, including that at Ephesus (37°56′27.75″N, 27°20′33.19″E) and at Aspendos. The former is Greek, the latter is Roman. Some regard the Aspendos theatre, built in 155 AD, as being the best preserved theatre of antiquity (Fig. 9.7).

Indeed, the Romans constructed many theatres. For example, the ancient theatre of Sabratha (Fig. 9.8) lies on the Mediterranean coast of Tripolitania (Libya) (Kenrick 2009). It is built from locally quarried sandstone. The date of construction, which is not known with complete certainty, can be placed at c 175–200 AD. This means it was started by Emperor Marcus Aurelius, continued by his son Commodus and completed with a lavish *proscaenium* by Emperor Septimus Severus. It had c 5000 seats. This theatre, unlike most other ancient theatres, which take advantage of a hillside to support their structure, was built on flat ground, which necessitated the construction of a substantial frame to form the *cavea* (seating area*)*. The magnificent stage backdrop is 25 m high, consisting of 108 Corinthian columns arranged in three stories. The theatre appears so complete because of its reconstruction by Italian archaeologists in the 1930s. After its reconstruction, the theatre was re-inaugurated by a letter day emperor, Benito Mussolini. The site was listed as a UNESCO World Heritage Site in 1982 (https://whc.unesco.org/en/list/184/) (accessed 4 April 2025). There is also a Roman theatre at neighbouring Leptis Magna.

At Cartagena on the coast of southeast Spain, there is a Roman theatre built into a hillside (Fig. 9.9). It was constructed between 5 and 1 BC, as has been proven by the dedication of the edifice to Gaius and Lucius Caesar. It had a capacity of some 6,000 spectators.

Finally, at Busra in Syria there is a splendid Roman theatre (Fig. 9.10). This was built of basalt in either the second century AD, probably during the reign of Trajan. The theatre is 102 m across and had seating for c 17,000 people. It is a World Heritage Site (https://whc.unesco.org/en/list/22/) (accessed 4 April 2025). It may have suffered some damage during recent internal strife in Syria (Al Kassem et al. 2024).

9.4 Hippodromes

The Greeks and Romans found pleasure not only in theatres and amphitheatres, but also in hippodromes, where horse and chariot races were held. Examples include the great Hippodrome in Byzantine Constantinople and the Circus Maximus in Rome itself (Fig. 9.11). In shape hippodromes were oblong, with one end semi-circular and the other square. In other words, they resembled a U with a closed top. Seats ran in

Fig. 9.5 Top: The Theatre at Siracusa, Sicily, Italy. 37° 4′33.46″N, 15°16′30.50″E. Scale bar = 70 m, ©Google Earth Bottom: From the inside (*Photo* A.S. Goudie)

tiers the length of the arena and along the curve, while at the straight end dignitaries occupied seats above the arena's offices. A low wall, called a *spina*, ran most of the length of the stadium and divided the course. This median strip was decorated with monuments and had sculptures. Because as many as 10 chariots raced at any one time, the breadth of the course was sometimes as much as 120 m; the length was typically about 180 to 210 m.

The Circus Maximus (Ames 2016) was larger than this, however. It reached a maximum size under Constantine (fourth century AD) of about 610 by 190 m, with a seating capacity of between 150,000 and 250,000 spectators, greater than that of any subsequent stadium.

Other notable Roman hippodromes include that at Jerash in Jordon, which was built sometime between the first and third centuries AD. Smaller than the Circus Maximus, this stadium (244m by 50m) was once surrounded by seating for up to 15,000 spectators (Fig. 9.12). Somewhat larger, the Roman hippodrome at Tyre in Lebanon is also well preserved (Fig. 9.13). Dating back to the second century, it is 480 m long and 90 mm wide. It may have been able to seat 20,000 spectators. Humphrey (1985) provides a detailed discussion of Roman circuses, including that at Leptis Magna in Tripolitania (32°37′59.29″N. 14°18′39.38″E).

Another type of sporting venue popular in ancient Greece was the gymnasium (Romano 2016). The name comes from the ancient Greek term *gymnós*, meaning 'naked' or 'nude'. Only adult male citizens were allowed to use the gymnasia. Associated with the gymnasia were stadiums and a *palaistra*. This essentially consisted of a rectangular court surrounded by colonnades with adjoining rooms. These rooms might house a variety of functions: bathing; ball playing; undressing and storage of clothes; seating for socializing and philosophizing, observation, or instruction; and storage of body oils, dust, or athletic equipment. Combat events such as boxing and wrestling were a speciality of the Greek 'palaistra', whereas the Roman 'palaestra' was an exercise area attached to public baths. The typical sports practised in a gymnasium were

Fig. 9.6 Top: Taormina, Sicily, Italy. 37°51′8.49″N, 15°17′32.18″E. Scale bar = 60 m, ©Google Earth. Bottom: Taormina Theatre from the inside (*Photo* A.S. Goudie)

Fig. 9.7 Roman Theatre at Aspendos in Türkiye. 36°56′20.06″N, 31°10′20.45″E. Scale bar = 50 m, ©Google Earth

Fig. 9.8 Top: Sabratha, Libya. 32°48′19.22″N, 12°29′7.00″E. Scale bar = 40 m, ©Google Earth. Bottom: Sabratha Theatre Stage (*Photo* A.S. Goudie)

wrestling, running, boxing, jumping, discus, and gymnastics. There were also sports that were useful for warfare, including archery, javelin, armed combat, and the use of catapults. Relationships could develop in the gymnasia between a young boy (*eromenos or catamite*) and a young man (*erastes or pederast*) (Fig. 9.14).

An excellent example of an ancient Greek gymnasium complex is that at Messini in the Peloponnese of Greece (Fig. 9.15), with its stadium and palaestra. This imposing group pf buildings was constructed during Hellenistic and late Roman times (third century BC–fourth century AD). It is on the World Heritage Site tentative list (https://whc.unesco.org/en/tentativelists/5859/) (accessed 4 April 2025).

The Greeks built bath houses (as at Olympia), but the Romans were perhaps even greater ones for bathing. They were responsible for the construction of sometimes elaborate bath house complexes, as at Caracalla, Rome (Fig. 9.16). Located close to the Circus Maximus, construction of these baths was probably initiated by Emperor Septimus Severus and completed during the reign of his son, Caracalla. The bath complex covered approximately 25 ha and is rectangular in shape, measuring 337 m × 328 m. It was inaugurated in AD 216. There were hot, tepid, and cold rooms; an Olympic-sized swimming pool; libraries; and a gymnasium. A new aqueduct, the Aqua Antoniniana (a branch of the earlier Aqua Marcia), was constructed to transport water to the baths.

Fig. 9.9 Top: Cartagena Roman Theatre, Spain. 37°35′57.77″N, 0°59′2.25″W. Scale bar = 80 m, ©Google Earth. Bottom: Ground image (*Photo* Alice Goudie)

Fig. 9.10 Roman Theatre at Busra, Syria. 32°31′3.49″N, 36°28′54.67″E. Scale bar = 80 m, ©Google Earth

Fig. 9.11 Circus Maximus in Rome. 41°53′10.14″N, 12°29′10.69″E. Scale bar = 200 m, ©Google Earth

Fig. 9.12 Roman Hippodrome at Jerash in Jordan. 32°16′25.14″N, 35°53′27.41″E. Scale bar = 100 m, ©Google Earth

Fig. 9.13 Roman Hippodrome at Tyre, Lebanon. 35°12′36.84″E, 35°12′36.84″E. Scale bar = 200 m, ©Google Earth

Fig. 9.14 Eromenos (right, with hoop) and Erastes (left) in the gymnasium (https://commons.wikimedia.org/wiki/File:Courtship_-_Bowl_-_Ancient_Greek_-_Athens_National_museum.jpg) (accessed 4 April 2025)

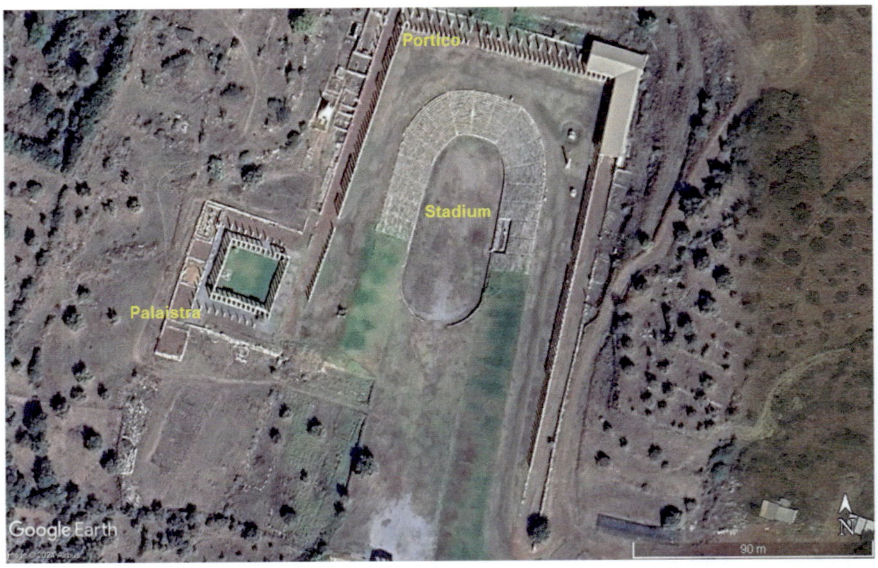

Fig. 9.15 Gymnasium and Wrestling School (Palaistra) at Messini, Greece. 37°10′21.94″N, 21°55′8.73″E. Scale bar = 90 m, ©Google Earth

Fig. 9.16 Caracalla Roman Baths, Italy. 41°52′44.24″N, 12°29′32.68″E, Scale bar = 100 m, ©Google Earth

References

Al Kassem A, Hopper K, Ibraheem IM, Hajj HE, Maier A, Ali AA, Richter J (2024) Assessment of the impact of the Syrian conflict on archaeological sites in the Daraa region. Levant. 1–22. https://doi.org/10.1080/00758914.2024.2420409. Accessed 4th April 2025

Ames CS (2016) Rome's seat of passion: an assessment of the archeology and history of the Circus Maximus. Cogent Arts Humanities 3(1):1168906

Aslanidis K (2003) The Roman odeion at Epidaurus. J Roman Archaeol 16:300–311

Frere SS, St Joseph JKS (1983) Roman Britain from the air. Cambridge Univ Press, Cambridge, p 232

Chase RG (2002) Ancient Hellenistic and Roman amphitheatres, stadiums and theatres. Univ Press of New England, Lebanon, New Hampshire, p 662

Harris R, Koser K (2018) Thysdrus and Ancient Rome. In: Harris R, Koser K (eds) Continuity and change in the Tunisian Sahel. Routledge, London, pp 91–97

Historic England (2018) Roman amphitheatres, theatres and circuses: Introductions to heritage assets. Historic England, Swindon. https://www.historicengland.org.uk/listing/selection-criteria/scheduling-selection/ihasarchaeology/. Accessed 4th April 2025

Humphrey JH (1985) Roman circuses: arenas for chariot racing. Univ California Press, Berkeley, Berkeley, p 703

Kenrick P (2009) Libya archaeological guides: Tripolitania. Society for Libyan Studies, London, p 224

Romano DG (2016) Athletics: stadia, gymnasia, palaistrai, and hippodromes. In: Miles MM (ed) A companion to Greek architecture. John Wiley & Sons, New York, pp 3314–3327

Royal Commission on Historical Monuments (England) (1970) An inventory of the historical monuments in the County of Dorset, Vol 2, part 3. HMSO, London and Woking, p 701

Sear F (1996) The theatre at Taormina—a new chronology. Papers Brit School Rome 64:41–79

Rituals and Religion: Churches, Temples, Pagodas, and Mosques

10

Abstract

Ancient religions and rituals have produced a whole suite of different types of buildings. These include the Neolithic henges and causewayed enclosures of England, ziggurats of the Middle East, Egyptian, Zoroastrian and Hindu temples, churches, mosques, and Buddhist pagodas. In some cases, such as Hagia Sophia in Istanbul, or the Cathedral in Siracusa (Sicily), they have been used by different religions during their long histories.

Keywords

Henge · Church · Temple · Pagoda · Mosque · Cathedral · Cemeteries

10.1 Introduction

Over the millennia, ancient rituals and religions have produced a whole suite of different types of buildings. These include the Neolithic henges of England ziggurats of the Middle East, Egyptian, Zoroastrian and Hindu temples, and Buddhist pagodas, churches, and mosques. In some cases, such as Hagia Sophia in Istanbul, or the Cathedral in Siracusa (Sicily), they have been used by different religions during their long histories and this is evident in their fabrics.

10.2 Mounds, Enclosures, and Henges of Britain

In southern England there is a landscape with a remarkable group of prehistoric sites that include mounds, henges, causewayed enclosures, and cursuses. The last of these, enclosure structures comprising parallel banks with external ditches or trenches, often appear as crop marks, but in general do not show up well on GE.

The largest Neolithic mound in Europe is Silbury Hill (Fig. 10.1) in Wiltshire, which is 40 m high and 160 m in diameter (Bayliss et al. 2007). It has a volume of 350,000 cubic metres. It was constructed in phases between 2700 and 2300 BC. Its actual purpose is far from clear (Atkinson 1969, 1970) in spite of a long history of excavation (Leary and Field 2010). There seems to be no evidence of any human burials in the mound, and its purpose may have been ceremonial. It certainly provided a panoramic view over a great ritual landscape.

A causewayed enclosure is a type of large prehistoric earthwork common to the early Neolithic in Europe (Oswald et al. 2014) (https://historicengland.org.uk/images-books/publications/iha-causewayed-enclosures/heag200-causewayed-enclosures/) (accessed 8 June 2023). It is an enclosure marked out by ditches and banks, with a number of causeways crossing the ditches. Also known as 'causewayed camps' or 'interrupted-ditch enclosures', they represent the earliest known examples of the enclosure of open space. They date to the early Neolithic (4000–3300 BC). Over 80 causewayed enclosures are now known in the British Isles and the majority are found in England south of the River Trent. More than 100 examples are recorded in France and further sites are known from Scandinavia, Belgium, Germany, Italy, and Slovakia. Unlike hill forts, they were not primarily defensive structures, nor were they funerary monuments. Some appear to have been fortified, others to have been used for ceremonial, ritual, market, and social purposes, but their functions still remain obscure. Figure 10.2 shows a causewayed enclosure at The Trundle near Chichester in southern England. This is sited near Goodwood Racecourse on the western part of the Sussex Downs in southern England. It is of Neolithic age, and its ditches enclose an area c 122 m across.

A. Goudie, *Landscape Heritage of the World with Google Earth*, https://doi.org/10.1007/978-3-031-95555-6_10

Fig. 10.1 Silbury Hill, Wiltshire, England. 51°24′56.21"N, 1°51′25.87"W. Scale bar = 100 m, ©Google Earth

Fig. 10.2 The Trundle, Causewayed Camp, West Sussex, England. 50°53′32.81"N, 0°45′13.00"W. Scale bar = 100 m, ©Google Earth

Dotted across Britain and Ireland there are a number of landscape features called henges. A henge is a prehistoric circular or oval earthen enclosure, dating from around 3000 BC to 200 BC, i.e. from the Neolithic to the early Bronze Age (https://www.english-heritage.org.uk/visit/inspire-me/what-is-a-henge/) (accessed 4 April 2025). There are fewer than 100 of them still surviving. Their key feature is a ring-shaped bank on the outside and a ring-shaped ditch on the inside. Some henges have multiple rings of bank-and-ditch. Because the ditch sits inside the bank, it is unlikely that they were defensive structures. Unlike hillforts, they were not necessarily built on high ground. Rather, it is thought, they were spaces for ritual or ceremonial performances. The barrier of the earthworks was more likely symbolic than functional. There are many different forms of henges and they may have had many different roles (Pollard 2012). They may have standing stones associated with them, as is the case with Stonehenge itself. Burl (1969) provides a map of their distribution and discusses their different forms.

The largest of the Wessex henges is that at Avebury (Fig. 10.3) (Burl 1979; Gillings et al. 2019). Avebury is part of an extraordinary set of Neolithic and Bronze Age ceremonial sites that seemingly formed a vast sacred landscape comprising the Stonehenge and Avebury World Heritage Site (https://whc.unesco.org/en/list/373/) (accessed 4 April 2025). They include West Kennet Avenue, West Kennet Long Barrow, The Sanctuary, Windmill Hill, and the enigmatic Silbury Hill. Like Stonehenge, Avebury has impressive standing stones. It is set on a flat plain east of the River Kennet, has a circumference of 1.3 km, an external diameter of 427 m and four entrances, all used by modern roads. It was probably built between 2900 and 2600 BC. The encircling bank is up

Fig. 10.3 Top: Avebury Stone Circle, Wiltshire, England. 51°25′42.92"N, 1°51′12.57"W. Scale bar = 200 m, ©Google Earth. Bottom: Avebury Stone Circle drawn by William Stukeley in 1724. How little it has changed

to 5.5 m high and between 23 and 30 m wide at the base. The ditch is still over 5 m deep and over 20 m wide at the top. Within the henge there is a ring of 98 stones standing just inside the inner lip of the ditch. The stones, composed of sarsens, are typically 10–11 m apart. The outer stone circle is the largest prehistoric circle of standing stones in the world. There are also other standing stones within this main circle. In the 1720s the antiquary William Stukeley drew and mapped the site.

Another interesting Wessex henge is that at Knowlton in Dorset (Fig. 10.4) (Gale 2017). Here there is a Neolithic henge—the Church Henge—with a 2 m high ditch and a 2 m high bank. It and its neighbours probably date to around the middle of the third millennium BC. The enclosure is orientated roughly northeast to southwest and measures 106 m by 94 m. Within the henge is a church that was first built in the twelfth century, had a tower added in the fourteenth century, and was then abandoned in the eighteenth century when the tower fell in. It is in a landscape packed with heritage constructions, including tens of barrows and ring ditches, four other henges and the Dorset Cursus.

Henges are not restricted to southern England. Some excellent examples occur at Thornborough in North Yorkshire and show up well on GE (54°12′29.81"N, 1°33′43.31"W).

10.3 Ziggurats

In Iraq lies the Ziggurat of Ur (Fig. 10.5). Ziggurats were staged towers that were surmounted by shrines and which had access by a series of staircases. In the flat landscape of Lower Mesopotamia, where it is located, these shrines would

Fig. 10.4 Top: Knowlton Circle (Henge) and abandoned Norman Church, Dorset, England. 50°53′31.60"N, 1°58′2.91"W. Scale bar = 70 m, ©Google Earth. Bottom: Ground image from the edge of the henge. Ditch in foreground, rampart far left (*Photo* A.S. Goudie). A large yew tee is located on the far northeast of the site and is decked with ribbons

have been visible for many kilometres around. The ziggurat, which may have been over 30 m in height, was built of bricks in the Early Bronze Age by kings of the Ur III dynasty, who held sway from 2112 to 2004 BC. It is the finest surviving example of Sumerian religious architecture. It is roughly rectangular in plan, with its corners oriented on the points of the compass. The ziggurat and the temple on its top were built for the moon goddess Nanna, the divine patron of the city-state. In the sixth century BC, four more tiers were added to make it a seven-storey structure, but only the first two tiers now survive. In the 1920s and early 1930s, Sir Leonard Woolley carried out major excavations and revealed the monument in its entirety (Wooley 1933). It is now part of a World Heritage Site (https://whc.unesco.org/en/list/1481/) (accessed 4 April 2025).

10.4 Egyptian Temples

As we saw in Chap. 7, there were many splendid temples built in Ancient Egypt as at Karnak and Luxor (Schulz and Seidel 1998). One great temple was that of the goddess Hathor at Dendera (Fig. 10.6), which was started in the late Ptolemaic period in 54 BC and then continued under Tiberius and Nero. Although never fully completed, Dendera is the best preserved example of a late Egyptian temple complex. The entrance of the temple, with six massive columns surmounted with the head of Hathor, is considered among the most beautiful monuments in Egypt. Hathor was among the most important and popular deities in ancient Egypt. She was associated with music, joy, dance, and motherhood. The temple was connected by a channel to the nearby River Nile. Encircling

Fig. 10.5 Top: Ziggurat of Ur, Iraq. 30°57′45.75"N, 46° 6′11.32"E. Scale bar = 40 m, ©Google Earth. Bottom: Ziggurat of Ur. Leonard Woolley's Excavation In 1923–24, (https://www.penn.museum/sites/journal/1235/) (Accessed 4th April 2025). Public domain

the temple area is a 280 × 280 m mud brick wall that is 10 m thick. The site also has a 20 × 30 m Sacred Lake, lined with sandstone blocks. This was used for ritual purposes. In the late fifth century AD a Christian church was built on the site (Wildung 1997).

10.5 Greek Temples

The remains of Greek temples are widespread in the classical world. In western Sicily, there is a cluster of Greek temples at a site called Selinunte (Fig. 10.7). This ancient Greek metropolis was built in the seventh century BC on a promontory overlooking the sea, and over two and a half centuries became one of the world's richest and most powerful cities. Destroyed by the Carthaginians in 409 BC, it finally fell to the Romans around 250 BC, after which it rapidly declined. The archaeological site contains various great Doric temples, the earliest dating from 550 BC, with five centred on an acropolis. Temple E (the Temple of Hera) (Becker et al. 2020) was built towards the middle of the fifth century BC on top of the foundations of a more ancient building. With its columns over 10 m high, it is the best preserved of the temples but its present appearance is partly the result of some reconstruction using original material in the 1950s.

Three other well-preserved Greek temples of note occur at Paestum near Naples (Fig. 10.8) (Cozzolino et al. 2019). One of these is the Temple of Hera II. It was built in the Doric order around 460–450 BC and so is of broadly similar age to Selinunte. It measures 24.5 by 60 m, with 6 external columns along its shorter sides and 14 columns along its longer sides (counting the corner columns twice). These

Fig. 10.6 Top: Dendera, Egypt. 26° 8′31.42"N, 32°40′13.45"E. Scale bar = 100 m, ©Google Earth. Bottom: Temple of Hathor, Dendera (*Photo* A.S. Goudie).Note the mud brick wall on the right side

are 8.9 m high and are made of golden coloured travertine. While eighteenth-century historians assigned the Temple to Poseidon (or Neptune to the Romans), more recent research has shown that this temple was more likely devoted to Hera, the goddess of marriage, women, and family. It was listed as a World Heritage Site in 1998 (https://whc.unesco.org/en/documents/130940/) (accessed 4 April 2025).

10.6 Arabian Temple

In Yemen there is a temple complex in the Ma'rib area that is now a World Heritage Site (https://whc.unesco.org/en/list/1700/) (accessed 4 April 2025). The finest temple is that of Awwam (Awam), which is also called Mahram Bilqis (Harrell 2007). The temple was in operation from the beginning of the first millennium BC until the fourth century AD. It is a testimony to the affluence and importance of the Kingdom of Saba, which dominated South Arabia in between the eighth century BC and the third century AD. It was the most important site of pilgrimage in pre-Islamic Yemen, centred around the Moon (?) god Almaqah. The building of the temple's massive enclosure by Mukarrib Yada 'il Dharih 1 took place in the middle of the seventh century BC. The enclosure is defined by a massive oval-shaped wall that is c 757 m long. A seventh century BC cemetery is attached to the Oval Sanctuary and around 20,000 estimated burials took place during its long period of usage (Fig. 10.9). Just to the west is another Sabean temple—Barran.

10.7 Cathedrals, Churches, and Cemeteries

The Cathedral in Siracusa (Fig. 10.10), Sicily, is another religious building with a complex history, showing the impact of various different faiths—Greek, Christian, and Islamic. It stands in the city's historic core on Ortygia Island. Its structure is originally a Greek Doric temple—the Temple of Athena. This was built in the sixth century BC (590–580 BC). The temple had six columns on the short sides and 14 on the

Fig. 10.7 Top: Selinunte Greek Temple, Sicily. 37°35′12.37"N, 12°50′5.63"E. Scale bar = 60 m, ©Google Earth. Bottom: Selinunte Temple E. (*Photo* A.S. Goudie)

long, and these can be seen both on the inside and on the outside of the cathedral. The building became a Christian church during the fifth century AD, as a result of the political decisions taken by Justinian to affirm Christianity. This was done by walling up the spaces between the Greek columns and cutting eight arches into the inner walls of the *cella* (the inner chamber of the temple where the image of the deity was housed). This created a basilica plan with a central nave and two side-aisles with the entrance to the West—at the opposite end to the main facade of the Greek temple which faced east. The building was converted into a mosque in 878 AD, then converted back again when Roger 1 of Sicily, a Norman, retook the city in 1085 AD. A dramatic Baroque façade was added in the eighteenth century. It is a veritable palimpsest (Van Ooijen 2019).

In Sardinia there is a small church of San Giovanni at Tharros. Though by no means as imposing as the cathedral in Siracusa, it is a superb example of an early and modest Christian Church (Fig. 10.11). Built around the sixth century AD during the Byzantine period, the church was constructed using sandstone from Roman and Punic ruins. It was modified in the eleventh century although elements of the sixth-century Byzantine original remain, including the characteristic red dome. It has a rectangular layout with a protruding apse oriented to the east. The rectangular hall is divided into three naves, covered by barrel vaults and ending in a similarly vaulted transept. In the central nave and the apse, there is a small dome supported by large pillars.

In India there are the remains of some hundreds of cemeteries and thousands of graves that were constructed during the time of the British Raj. Many occur in the old military cantonment areas in places like Karnal and Amballa, and others occur in the great cities such as Kolkata and Chennai (Fig. 10.12).

One of the old English cemeteries is at Surat in Gujarat, Western India (Scarre and Roberts 2005). This is indeed the earliest of the cemeteries founded by the traders of the East India Company in South Asia and it provides one of

Fig. 10.8 Top: Temple of Hera II at Paestum, Italy. 40°25′11.87"N, 15° 0′19.33"E. Scale bar = 40 m, ©Google Earth. Bottom: Temple of Hera II, Paestum, from the west (*Photo* A.S Goudie)

Fig. 10.9 The Temple of Awwam, Yemen. 15°24′14.30"N, 45°21′21.22"E. Scale bar = 80 m, ©Google Earth

Fig. 10.10 Top: The Cathedral, Siracusa, Sicily, Italy. 37° 3′33.82"N, 15°17′37.55"E. Scale bar = 40 m, ©Google Earth. Bottom: Siracusa Cathedral, showing the Doric columns of the Greek Temple in the background (*Photo* A.S. Goudie)

the most visible reminders of the early British presence in the sub-continent. The city of Surat was the leading port of the west coast of India until the rise of Bombay during the eighteenth century. It was the leading port of the west coast of India until the rise of Bombay during the eighteenth century. In 1607, the East India Company attempted to establish a 'factory' at Surat and the cemetery appears to have been established in 1649. It contains seventeen substantial monuments (leaving aside simple table tombs) and all of these monuments are constructed of brick covered with a thick surface of yellow stucco. The purpose of these large and elaborate monumental tombs was clearly to impress and to project an image of wealth and status. The style of these early tombs was distinctively Indo-Islamic.

In England, during the reign of Henry VIII, when the monarchy, following the Reformation, espoused Protestantism, hundreds of Roman Catholic religious buildings were destroyed. Their ruins are widespread. One of these is Hailes Abbey in Gloucestershire, details of which are given in (https://www.english-heritage.org.uk/siteassets/home/visit/places-to-visit/hailes-abbey/history/hailes-abbey-plan.pdf) (accessed 24 March 2025) (Fig. 10.13). In 1242, Richard, Earl of Cornwall (1209–72), who was the second son of King John, vowed to found a monastery to give thanks to God after surviving a storm at sea. Three years later, he was able to fulfil this vow when his brother Henry III granted him the manor of Hailes on the slopes of the Cotswold Hills. Richard decided that his new monastery should belong to the Cistercian order of monks (Carter 2018). In 1270, Richard's son Edmund presented the monks with what was believed to be a portion of the blood shed by Christ on the Cross. In 1277, the east end of the church was rebuilt to house this holiest of all relics. The Holy Blood of Hailes turned the abbey into a highly important and much-visited pilgrimage destination. However, in the early 1530 s Henry VIII sought to reform what he viewed as the abuses of the Catholic Church and launched attacks on the veneration of relics. By

Fig. 10.11 Top: The Church of San Giovanni at Tharros, Sardinia, Italy. 39°52′56.37"N, 8°26′19.46"E. Scale bar = 30 m, ©Google Earth. Bottom: Exterior view, showing the three naves and the red dome (*Photo* A.S. Goudie)

1538 the relic had been denounced as a fake and the shrine destroyed. The following year Hailes fell victim to King Henry VIII's Suppression of the Monasteries, and its end came on Christmas Eve 1539 when the last Abbot and the community of 21 monks signed the surrender deed. The site was pillaged by the locals.

10.8 Hagia Sophia

Hagia Sophia (Fig. 10.14) is one of many gems in Istanbul, Türkiye. They form a World Heritage Site (Historic Areas of Istanbul) (https://whc.unesco.org/en/list/356/) (accessed 4 April 2025). In 532 AD Emperor Justinian commissioned Anthemius of Tralles (a mathematician) and his assistant Isidore of Miletus (an engineer) to create a new church of unprecedented grandeur. With its enormous dome, it remained for a thousand years the world's largest cathedral. In order to create the largest interior space possible, the two architects designed a colossal central dome and supported it using a revolutionary construction method called *pendentives*. Hagia Sophia has four of these which allow for the weight of the circular dome to transition to a square supporting superstructure below without either massive pillars or columns interrupting the internal space. The structure now standing is essentially the sixth-century edifice, although an earthquake caused a partial collapse of the dome in 558 (restored 562). In 1452 AD, after the Ottoman capture of what was then Constantinople, Sultan Mehmet II converted it to a mosque, and it was the main mosque of Istanbul until the 1616 construction of the nearby Sultan Ahmed Mosque. Islamic additions included four minarets, a *minbar* (pulpit) and a *mihrab* (a niche indicating the direction to Mecca), but much of the Christian fabric remained. In the 1930s it became a museum but once again became a mosque in 2020 (Ahmed and Ahmad 2021).

Fig. 10.12 English Tombs at Surat, India. Top: 21°12′45.56"N, 72°49′43.04"E. Scale bar = 40 m, ©Google Earth. Bottom: view in c 1855 (https://commons.wikimedia.org/wiki/File:Tombs_of_the_Early_English_Settlers,_Surat_(9185928036).jpg) (accessed 4 April 2025)

10.9 Jesuit Mission Stations

As the histories of Hailes Abbey, Hagia Sophia, and of the cathedral in Siracusa demonstrate, the fortunes of Christian churches have not always been easy. Another example is provided by the history of Jesuit missions in South America. Conceived by Basque former soldier Ignatius of Loyola, and officially established by Pope Paul III's bull *Regimini militantis ecclesiae* on September 27 1540, the Society of Jesus, or the Jesuit order, became one of the central instruments of the Catholic Church's counter-reformation. This was the case in South America where a group of former missions have been declared a World Heritage Site—Jesuit Missions of the Guaranis: San Ignacio Mini, Santa Ana, Nuestra Señora de Loreto and Santa Maria Mayor (Argentina), Ruins of São Miguel das Missoes (Brazil) (https://whc.unesco.org/en/list/275/) (accessed 4 April 2025). The missions were primarily to evangelize and 'civilize' the local populations.

San Ignacio Mini was one of the missions founded in 1610 in Argentina by the Jesuits during the Spanish colonial period. The original mission was erected around 1610, but did not settle in its current location until 1696. In the eighteenth century, the mission had a population of around 3000 people, mostly indigenous Guaranis. Shortly after the Suppression of the Society of Jesus in 1767, the Jesuits left and the mission buildings were destroyed by war in 1817. The remains of a church, cemetery, central square, priest's house, and a 200 further dwellings have been located, many of which were built of sandstone in a unique style known as 'Guaraní baroque' (Fig. 10.15 top).

Fig. 10.13 Top: Hailes Abbey, Gloucestershire, England. 51°58′6.79"N, 1°55′41.21"W. Scale bar = 70 m, ©Google Earth. Bottom: Hailes Abbey Cloister (https://upload.wikimedia.org/wikipedia/commons/f/f0/Hailes_Abbey_Landscape.jpg) (accessed 4 April 2025)

São Miguel das Missões in Rio Grande do Sul state, southern Brazil, was founded by the Jesuits in 1632, but in 1750, the Treaty of Madrid was signed between Portugal and Spain, which redivided sovereignty of certain parts of Brazil. The Guaraní converts refused to move village, which led to the mission being attacked by joint Portuguese and Spanish forces. The expulsion of the Jesuits from Brazil in 1759 further cemented its fate (Fig. 10.15 bottom).

10.10 Bagan

Bagan (Fig. 10.16) in Myanmar is an extraordinary assemblage of temples, stupas, and monasteries. It reached its peak between the eleventh and fourteenth centuries AD (Hudson et al. 2001). It is a World Heritage Site (https://whc.unesco.org/en/list/1588/) (accessed 4 April 2025). From the ninth to thirteenth centuries, the city was the capital of the Bagan Kingdom, the first kingdom that unified the regions that would later constitute Myanmar. During the kingdom's height, more than 10,000 Buddhist pagodas and monasteries were constructed in the Bagan plains alone, of which over 2200 survive. The city was once home to some 50,000 to 200,000 people, but declined after the Bagan Empire collapsed in 1287 due to repeated Mongol invasions. Since then, Bagan has suffered from floods and a number of earthquakes, notably in 1975 and 2016. It has also undergone a major reconstruction programme and Hudson (2008) reported that at least 1,299 Buddhist temples, monasteries, and stupas had been speculatively rebuilt from mounds of rubble since 1995. A further

Fig. 10.14 Top: Hagia Sophia, Istanbul, Türkiye. 41° 0′30.24″N, 28°58′48.95″E. Scale bar = 100 m, ©Google Earth. Bottom: Hagia Sophia in 1965 (*Photo* A.S. Goudie)

688 damaged buildings have received major repairs. The nature of the reconstruction is highly controversial. Mercifully, the March 2025 Myanmarh earthquake seems to have little effect.

10.11 Buddhist Sites

In Bangladesh, at Paharpur (Fig. 10.17), there is a great Buddhist monastery called Somapura Mahavira (Van Lohuizen-de Leeuw 1957). It is one of the most famous examples of architecture in pre-Islamic Bangladesh and was built by the second Pala king Dharmapala in c 781–821 AD. Enclosed by walls up to 5 m high, and made of baked brick, this monastery is quadrangular in form, and has a huge temple with a cross-shaped floor plan in the centre of the courtyard. The monastery also had cells for 177 monks. The site covers 11 ha. It was designated as a UNESCO World Heritage Site in 1985 (https://whc.unesco.org/en/list/322/) (accessed 4 April 2025).

In central Java, Indonesia, Borobudur (Fig. 10.18) is an outstanding Buddhist temple complex (Kowal 2019). It is said to be the largest Buddhist temple in the world. The structure was built between AD 750 and AD 842. Buried for centuries in forest, its rediscovery in the early nineteenth century was partly due to the British Lieutenant Governor of Java, Thomas Stamford Raffles. The temple was covered with soils and ashes from volcano eruptions and had gradually become overgrown with trees and shrubs when it was rediscovered.

It was built in three tiers: a pyramidal base with five concentric square terraces, the trunk of a cone with three circular platforms and, at the top, a monumental stupa. More than 55,000 cubic metres of andesite stone were carved to build this massive structure. Borobudur was constructed in the eighth century and subsequently abandoned following the fourteenth-century decline of Hindu kingdoms in Java and the Javanese conversion to Islam. It is a World Heritage Site (https://whc.unesco.org/en/list/592/) (accessed 4 April 2025).

Fig. 10.15 Top: San Ignacio Mini, Argentina. 27°15′18.33"S, 55°31′53.06"W. Scale bar = 80 m, ©Google Earth. Bottom: São Miguel das Missoes. 28°32′53.27"S, 54°33′18.31"W. Scale bar = 60 m, ©Google Earth

Another highly important Buddhist site is Anuradhapura in Sri Lanka (8°20′59.95"N, 80°23′47.74"E). This is also a World Heritage Site (https://whc.unesco.org/en/list/200/) (accessed 4 April 2025).

10.12 Hindu Hampi

The city of Hampi (Fig. 10.19) in southern India was part of the Hindu Kingdom of Vijayanagar which reached its peak under Krishna Deva Raya (1509–1530 AD) (Michell 2017). The temples of Ramachandra (1513 AD) and Hazara Rama (1520) are among the finest in India. Hazara Rama Temple is dedicated to Lord Rama, the Hindu deity. It was once the private temple of the kings and the royal family of Vijayanagara. Hampi was conquered by the Deccan Muslim confederacy in 1565. The city was pillaged before being abandoned. Hampi is a World Heritage Site (https://whc.unesco.org/en/list/241/) (accessed 4 April 2025). Another remarkable Hindu (and Jain) site in peninsular India, noted by tourists for its sometimes erotic and bestial sculptures, is Khajuraho (24°51′0.63"N, 79°55′10.14"E), but alas it does not as yet show up well on GE.

10.13 Zoroastria

Another important eastern religion, especially in Iran and India, is Zoroastrianism, a monotheistic faith established by the Persian prophet Zoroaster (Zarathustra) between c. 1500

Fig. 10.16 Top: Old Bagan Pagodas, Myanmar. 21° 9′58.41"N, 94°51′34.18"E. Scale bar = 80 m, ©Google Earth. Bottom: Old Bagan Pagodas (https://commons.wikimedia.org/wiki/File:Old_Bagan,_Myanmar,_Ancient_Buddhist_pagodas_and_temples_in_Bagan_plains.jpg) (accessed 4 April 2025). Public Domain

and 1000 BC. Takht-e Soleyman (Fig. 10.20) (The throne of Solomon) in northwest Iran is a Zoroastrian sanctuary with a fire temple (Azari-Dehkordi 2011). The site includes not only the principal sanctuary, which was partly rebuilt in the Ilkhanid (Mongol) period (thirteenth century), but also a temple of the Sasanian period (sixth and seventh centuries) dedicated to the goddess Anahita, the divinity of the waters. The core of the site is a fortified oval platform rising about 60 m above the surrounding plain and measuring about 350 m by 550 m. The complex is surrounded by a fortified mud brick wall interspersed with semi-circular bastions, and within this there is a large courtyard with an artificial lake at its centre. The lake itself is roughly 80 by 120 m and its overall depth averages about 70 m, but is 120 m at its deepest point. After the collapse of the Sasanian Dynasty caused by the Byzantine army of Heraclius I (the Roman Emperor) in 627 AD, Takht-e Soleyman was reduced to ruins. It is a World Heritage Site (https://whc.unesco.org/en/list/1077/) (accessed 4 April 2025).

Other Zoroastrian sites in the Kerman District of Iran, consisting of Towers of Silence, have been mapped using GE by Lohari et al. (2024). Towers of Silence are circular, raised structures built by Zoroastrians for excarnation. This is the exposure of human corpses to the elements, including vultures, with the purpose of enabling their decomposition. The Zoroastrian Towers of Silence of the Ardakan District of Yadz have also been illustrated using GE by Rahbar (2024) (Fig. 10.21). Their use was banned about 70 years ago.

Fig. 10.17 Paharpur Buddhist Monastery, Bangladesh. 25° 1′52.32"N, 88°58′38.28"E. Scale bar = 80 m, ©Google Earth

Fig. 10.18 Top: Borobudur Temple, Indonesia. 7°36′28.66"S, 110°12′14.18"E. Scale bar = 80 m, ©Google Earth. Bottom: Borobudur (https://commons.wikimedia.org/wiki/File:Pradaksina.jpg) (accessed 4 April 2025)

Fig. 10.19 Hazara Rama temple, Hampi, India. 15°19′4.22"N, 76°28′7.82"E. Scale bar = 50 m, ©Google Earth

Fig. 10.20 Takht-e Soleyman, Iran. 36°36′17.17"N, 47°14′6.28"E. Scale bar = 100 m, ©Google Earth

Fig. 10.21 Towers of Silence (Dakhma) of Ardakan, Yadz, Iran. 32°17′19.68"N, 54° 3′13.77"E. Scale bar = 40 m, ©Google Earth

Fig. 10.22 Top: Samarra Mosque, Iraq. 34°12′23.32"N, 43°52′48.52"E. Scale bar = 300 m, ©Google Earth. Bottom: Old picture of the mosque at Samarra with the Malwiya minaret in the foreground and Al-Askari shrine in the background (https://commons.wikimedia.org/wiki/File:Rare-pic-of-past-beautiful-in-Iraq-02_(cropped).jpg) (accessed 4 April 2025). Public domain

10.14 Islam

Finally, let us turn to some wonders of Islam. Samarra (Fig. 10.22), a former powerful Islamic centre in Iraq, was the capital of the Abbasid Empire (Northedge 2005). It stands on the east bank of the River Tigris 125 km north of Baghdad. The whole site is huge, being 41.5 km from north to south, and with a width varying from 4 to 8 km. Of special significance is the Great Mosque which was commissioned in 848 AD and completed in 851 AD by the Abassid caliph Al-Mutawakkil. At the time of construction, it was the world's largest mosque. The main building is encompassed by an outer baked brick wall supported by a total of 44 semi-circular towers including one at each corner. The Malwiya Tower is a spiralling cone, 52 m high and 33 m wide with a spiral ramp. It was constructed of sandstone. In 892 AD the caliph transferred the capital to Baghdad, causing Samarra's decline so that by 1300 most of the town was in ruins. Samarra is a World Heritage Site (https://whc.unesco.org/en/list/276/) (accessed 4 April 2025).

In Spain, there are many mosques that were constructed during the Islamic occupation, but which are now abandoned or ruined. One of these is the Aljama mosque in Medina Azahara near Cordoba (Fig. 10.23). The mosque building was divided between an open courtyard to the northwest and an interior prayer hall to the southeast. It is part of a World Heritage Site (https://whc.unesco.org/en/documents/168769) (accessed 4 April 2025).

The Medina was built in the mid-tenth century AD (c 940) by Abd ar-Rahman, the first caliph of Al-Andalus. It served as the capital of the Caliphate of Córdoba until it was sacked and pillaged by Berber factions in c 1010. It may also have been damaged by seismic activity (Rodríguez-Pascua et al. 2023).

Fig. 10.23 Top: The Caliphate Centre of Medina Azahara with the Aljama Mosque, Spain. 37°53′10.05"N, 4°52′1.64"W. Scale bar = 200 m, ©Google Earth. Bottom: Ground shot of Aljama Mosque from the northwest (https://commons.wikimedia.org/wiki/File:Cordoba_-_Medina_Azahara_08.jpg) (accessed 4 April 2025)

References

Ahmed A, Ahmad M (2021) Hagia Sophia: a study of its historical perspectives, architectural structure and geographical importance. VFAST Trans Islam Res 9(1):5–9

Atkinson RJC (1969) The date of Silbury Hill. Antiquity 43:216

Atkinson RJC (1970) Silbury Hill, 1969–70. Antiquity 44:313–314

Azari-Dehkordi F (2011) Introducing Geo-Cultural landscapes in Iran. In: Hong SK, Kim JE, Wu J, Nakagoshi N (eds) Landscape ecology in Asian cultures. Ecol Res Monogr, Springer, Tokyo. https://doi.org/10.1007/978-4-431-87799-8_5, pp 55–68 (Accessed 4th April 2025)

Bayliss A, McAvoy F, Whittle A (2007) The world recreated: redating Silbury Hill in its monumental landscape. Antiquity 81:26–53

Becker J, Miller JN, Canales E, Elliott T, Rabinowitz A (2020) Temple E at Selinus: a Pleiades place resource, Pleiades: A Gazetteer of past places. https://pleiades.stoa.org/places/502577470 (Accessed 4th April 2025)

Burl HAW (1969) Henges: internal features and regional groups. Archaeol J 126(1):1–28

Burl HAW (1979) Prehistoric avebury. Yale Univ Press, New Haven, p 275

Carter M (2018) Abbots and aristocrats: patronage, art and architecture at Hailes Abbey in the Late Middle Ages. J Brit Archaeol Ass 171(1):154–175

Cozzolino M, Longo F, Pizzano N, Rizzo ML, Voza O, Amato VA (2019) Multidisciplinary approach to the study of the temple of Athena in Poseidonia-Paestum (Southern Italy): New Geomorphological, geophysical and archaeological Data. Geosciences 9(8). https://doi.org/10.3390/geosciences9080324

Gale J (2017) Knowlton circles: a later neolithic and early Bronze Age ceremonial complex and its environs—a review. Landscapes 18(2):102–119

Gillings M, Pollard J, Strutt K (2019) The origins of Avebury. Antiquity 93:359–377

Harrell JA (2007) Building and ornamental stones of the Awam (Mahram Bilqis) Temple in Marib. Yemen. Arabian Archaeol Epigr 18(2):182–192

Hudson B (2008) Restoration and reconstruction of monuments at Bagan (Pagan), Myanmar (Burma), 1995–2008. World Archaeol 40(4):553–571

Hudson B, Lwin N, Maung W (2001) The origins of Bagan: new dates and old inhabitants. Asian Perspect 40:48–74

Kowal K (2019) The Borobudur temple: the Buddhist architecture in Indonesia. Budownictwo i Architektura 18(2):5–19

Leary J, Field D (2010) The story of Silbury Hill. Historic England, London, p 224

Lohari S, Mehrafarin R, Fathi H (2024) Studying the architecture of Kerman's Zoroastrian crypts with a look at the burial ritual. J Sistan Baluchistan Studies 4(2):25–40

Michell G (2017) Introducing Hampi: landscapes and legends. In: Mallik A, Chaudhury S, Chandru V, Srinivasan S (eds) Digital hampi: preserving Indian cultural heritage. Springer, Singapore. https://doi.org/10.1007/978-981-10-5738-0_3 (Accessed 4th April 2025)

Northedge A (2005) The historical topography of Samarra. British Institute for the Study of Iraq, British Academy, London, p 426

Van Ooijen JA (2019) Resilient matters: the cathedral of syracuse as an architectural palimpsest. Architect Hist 7(1). https://doi.org/10.5334/ah.65 (Accessed 4th April 2025)

Pollard J (2012) Living with sacred spaces: the henge monuments of Wessex. BAR Int Ser 2440:93–107

Rahbar M (2024) Zoroastrian Dakhma of Turkabad in Ardakan, Yazd Iran. AncT Iran Stud 3(8):57–83

Rodríguez-Pascua MÁ et al (2023) Archaeoseismological evidence of seismic damage at Medina Azahara (Córdoba, Spain) from the early 11th century. Applied Sci 13(3):1601

Scarre C, Roberts J (2005) The English cemetery at Surat: pre-colonial cultural encounters in western India. Antiquaries J 85:251–291

Schulz R, Seidel M (1998) Egypt. The world of the pharaohs. Kõnemann, Cologne, p 538

van Lohuizen-de Leeuw JE (1957) The ancient Buddhist monastery at Paharpur. Antiq Surviv 2:29–42

Wildung D (1997) Egypt from prehistory to the Romans. Taschen, Cologne, p 237

Woolley CL (1933) Excavations at Ur. J Roy Soc Arts 82:46–59

Geoglyphs

11

Abstract

Geoglyphs are large designs or motifs created on the ground by manipulating stones, gravel, or earth. There are many examples from around the world, and in this section we show some from Arabia, southern England, Peru and Chile, Kazakhstan, and the USA. The reasons for producing geoglyphs are multiple, and sometimes far from clear. Some may have been for herding and capturing animals, some may have had ritual significance, and some may have been boundary and route markers.

Keywords

Geoglyph · Nazca lines · Jordanian kites · Intaglios · Amazonia

11.1 Introduction

Geoglyphs are large designs or motifs created on the ground by manipulating stones, gravel, or earth. There are many examples from around the world, and in this section we show some from Arabia, southern England, Peru and Chile, and the USA. They are notoriously difficult to date, and their purposes are often far from clear. A good global summary is provided by Valenzuela and Clarkson (2020). GooglE was used to locate geoglyphs in the Thar Desert of Rajasthan, India (Oetheimer and Oetheimer 2021) and the Konkan region of Maharashtra, India (Sukumaran et al. 2025).

11.2 The Middle East Wheels and Kites

In the boulder-strewn basalt deserts of northeast Jordan, notably near Azraq (Fig. 11.1), there are hundreds of intriguing largely circular patterns (Rollefson et al. 2016). Some have the form of a wheel (with spokes radiating from the centre to a rim) while others are a collection of irregular enclosures inside a circle and have been likened to jellyfish. The rings may be c 50 m in diameter, and the circular huts inside are c 4–5 m in diameter. They may represent the regularly visited camping grounds of nomads, perhaps extending back as far as sixth millennium BC. The stone walls are low—now less than a metre—and may have been animal pens, while the rings may be foundations of houses (Kennedy and Bewley 2004, p. 75).

The so-called Jordanian 'kites' (Fig. 11.2) are best seen from the air. There are thousands of these stone-built structures in Arabia (Repper et al. 2022; Grocutt and Carleton 2021), and they have dimensions from 100 m upwards. They may occur in great swarms (Helms and Betts 1987). Crassard et al. (2015) provide a map of their distribution. The tails of the kites generally open out along the bottom of a slope and then converge before entering the head, which lies over the crest out of sight (Kennedy and Bewley 2004, p. 70–72). The heads can be star-shaped. It is thought that they were animal traps for trapping migrating gazelle, ostrich, and the like. They became redundant with the arrival of hunters with guns and motor vehicles. The use of some for controlling and herding domesticated flocks cannot be completely dismissed (Swizzero and Tisdell 2018). Their age is still debated, but Betts and Burke (2015) suggest they may, in some cases, date back to the Neolithic. Broadly similar features occur near Aswan in Egypt (Storemyr 2011), in Armenia (Malkinson et al. 2018), in the Ustyurt Plateau to the east of the Caspian Sea (Barge et al. 2016), in Tripolitania and the north Sahara (Giannelli and Maestrucci 2018, 2019), and in the Keimoes area of northwestern South Africa (Lombard et al. 2021), where they may have been used to trap springbok. Certainly, as the above-cited articles demonstrate, GE has been immensely important for mapping its distribution and analysing its varied forms (Kempe and Al-Malabeh 2013).

A. Goudie, *Landscape Heritage of the World with Google Earth*, https://doi.org/10.1007/978-3-031-95555-6_11

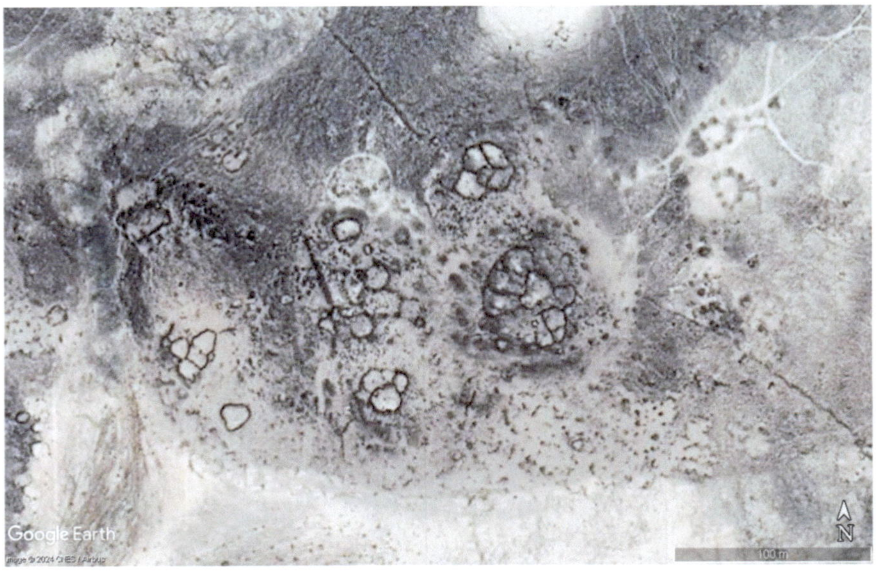

Fig. 11.1 'Wheels' at Azraq, Jordan. 32° 3′46.54"N, 37°12′13.85"E. Scale bar = 100 m, ©Google Earth

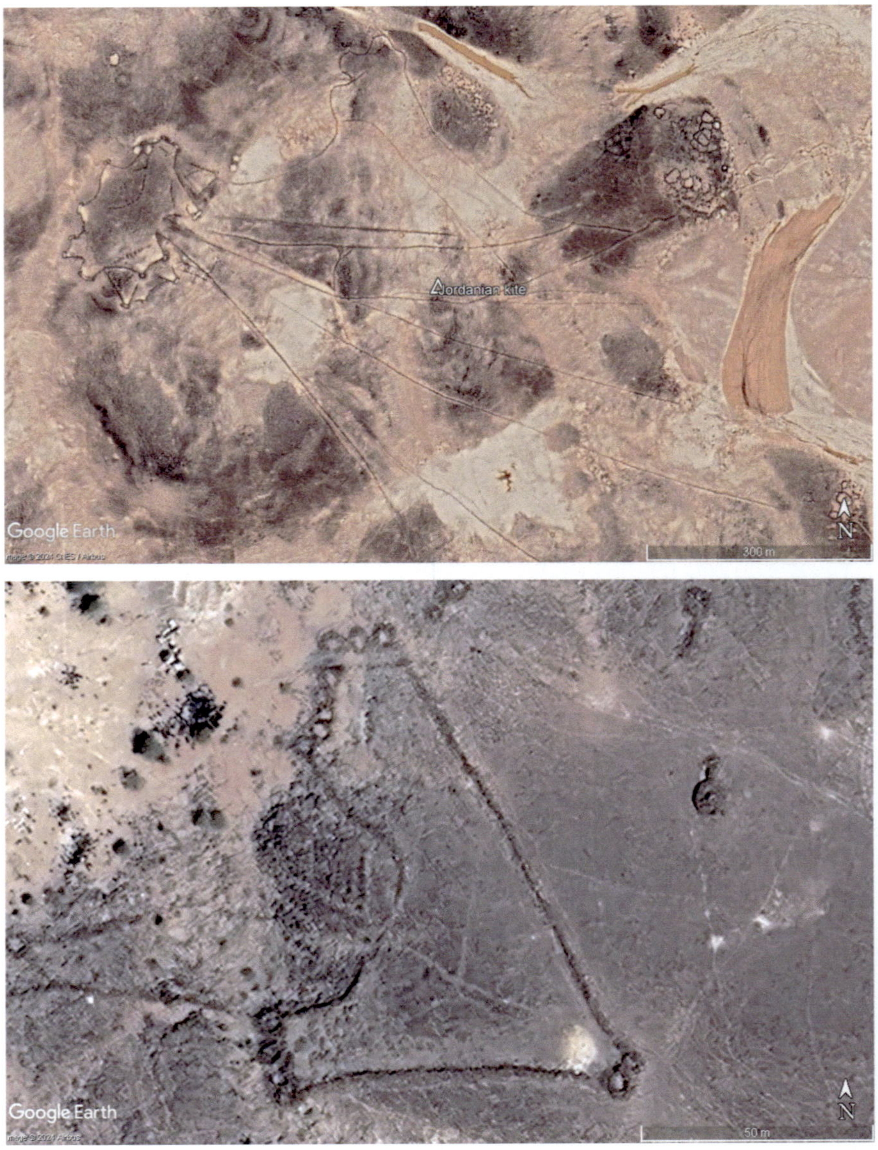

Fig. 11.2 Top: Jordanian 'Kites', Jordan. 32° 1′3.74"N, 37° 2′21.81"E. Scale bar = 300 m, ©Google Earth. Bottom: Prehistoric Animal Traps, Khaybar, Saudi Arabia, 25°41′59.02"N, 39°18′16.88"E. Scale bar = 50 m, ©Google Earth

11.3 The English Chalklands

One hundred or more figures have been cut into the white Chalk in various parts of southern Britain. Some are modern, but some are of great antiquity.

At Cerne Abbas in Dorset, southern England, there is a 55 m long figure carved into a hill (Fig. 11.3). This figure, the Cerne Giant, is a naked male wielding a large club. The overall form is naturalistic, his nipples, eyes, eyebrows, and an 11 m erect penis and scrotum being boldly represented. The date is uncertain. It is referred to in literature in the seventeenth century, but is almost certainly older. Indeed, recent optically stimulated luminescence dating has suggested an origin between the years 700 AD and 1110 AD, and research shows that he was originally carved as an image of the Classical hero Hercules as a rallying point for mustering West Saxon armies at a time when Dorset was being attacked by Viking armies (Papworth 2021; Gittos and Morcom 2024).

The Uffington White Horse (Fig. 11.4), incised on the chalk downs of south Oxfordshire in England, is a remarkable beast. As Abram (2022, p. 184) wrote, it 'is the oldest surviving hill figure in Britain and, quite simply, one of the great masterpieces of prehistoric art in Europe. Its sweeping lines, spread over 110 m (360 feet) from head to tail, convey with Piccasso-like minimalism the feel of a steed in motion, frozen in mid-stride with its body outstretched'. Recent dating has placed the creation of the work to the very end of the Bronze Age at 1000–800 BC (Miles et al. 2014). It may have had a ritual or religious purpose. Alternatively it could possibly be a tribal totem marking a land boundary. Also, the argument has now be made that the Uffington carving is a representation of the sun-horse found in iconography throughout later prehistoric Europe (Pollard 2017).

Lying close to the White Horse is The Ridgeway, sometimes called Britain's oldest road. This is a track that runs from Wiltshire along a chalk ridge of the Berkshire Downs to the River Thames, and then along the Icknield Way, which stretches from Salisbury Plain to East Anglia on the east coast. Evidence of human activity along the Ridgeway dates back to the Neolithic and it linked several important sites, such as the Avebury stone circles and the long barrows at Wayland 's Smithy. During the Bronze Age and Iron Age, the Ridgeway served as a vital artery, connecting settlements and hillforts such as Uffington Castle (https://www.historytools.org/stories/the-ridgeway-a-journey-through-time-on-britains-oldest-road) (Accessed 4th April 2025).

11.4 South America

The Nazca Lines (Fig. 11.5) occur in the Atacama Desert in Peru about 350 km southeast of the modern capital of Lima. They were created by the Nazca culture which flourished from 200 BC to 500 AD. They were created by scraping the dark, patinated stones of the desert pavement to reveal the underlying lighter-coloured, unvarnished sediment beneath. More than 640 km^2 of the desert are covered with these lines, and they have a combined length of 1300 km. They comprise a range of forms: recognisable figures such as animals (including the hummingbird, a condor, a monkey, a spider, and a whale), geometrical shapes, and seemingly random lines (Jones 2020, p. 68–69). They are a World Heritage Site (https://whc.unesco.org/en/list/700/) (Accessed 4th April 2025) and may have had ritual astronomical significance, though this is a matter of huge debate. The World Heritage

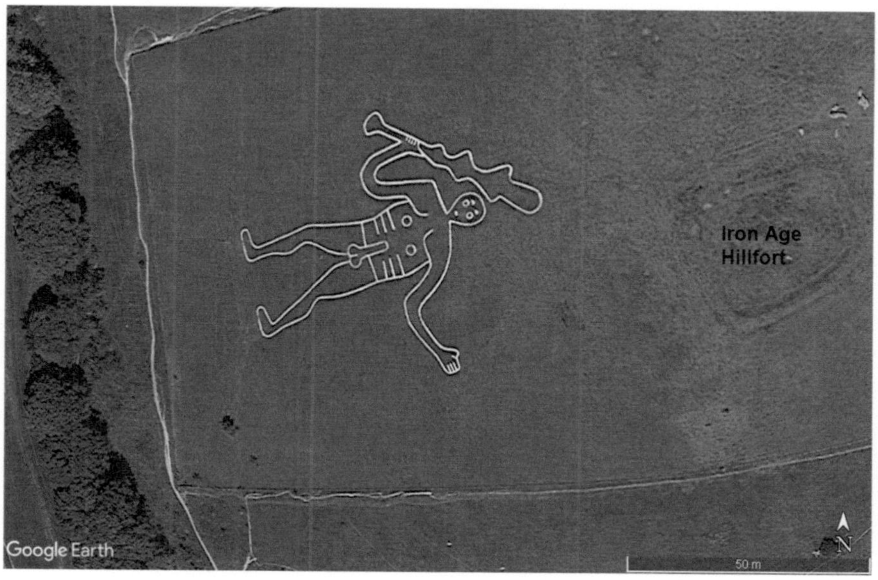

Fig. 11.3 Cerne Abbas Giant, Dorset, England. 50°48′49.24"N, 2°28′29.12"W. Scale bar = 50 m, ©Google Earth

Fig. 11.4 Top: Uffington White Horse, Oxfordshire, England. 51°34′39.72"N. Scale bar = 40 m, ©Google Earth. Bottom, the nearby Ridgeway. 51°34′32.10"N, 1°34′9.03"W. Scale bar = 300 m, ©Google Earth

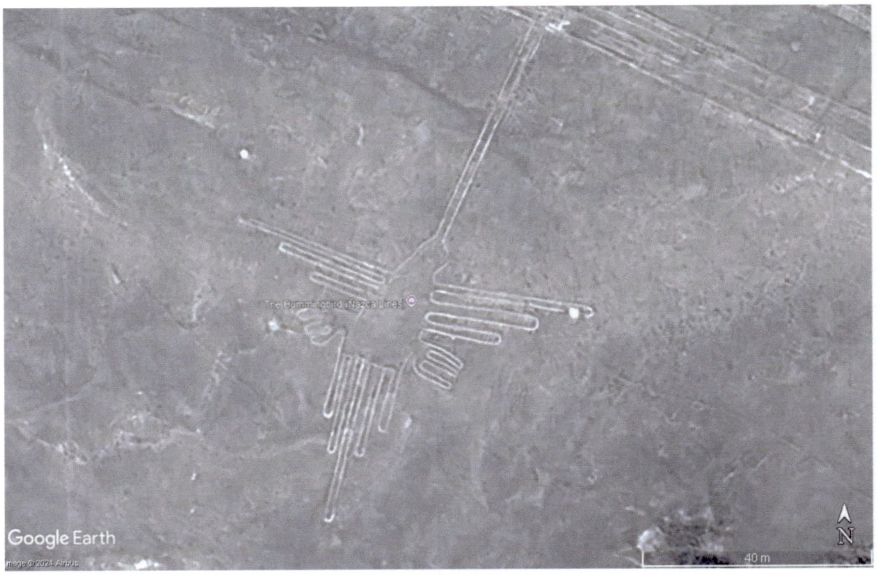

Fig. 11.5 Hummingbird, Nazca Lines, Peru. 14°41′31.56"S, 75° 8′55.71"W. Scale bar = 40 m, ©Google Earth

nomination remarks that 'They are the most outstanding group of geoglyphs anywhere in the world and are unmatched in its extent, magnitude, quantity, size, diversity and ancient tradition to any similar work in the world'. Their mapping and classification are discussed by Sakai et al. (2024), while damage to them is discussed in Sect. 1.3.

The Paracas Candelabra (Fig. 11.6), also El Candelabro (the Trident), is a prehistoric geoglyph, dated to around 200 BC, found on the northern face of the Paracas Peninsula at Pisco Bay in Peru. The design is cut 0.6 m into the desert surface, with stone walls. With a large, branch-like appearance, the purpose and meaning of the Candelabra remain unknown. The shape of the geoglyph is the subject of speculation. Some have likened it to a cactus plant, while others believe it looks more like a three-branched candlestick, hence the name 'candelabra'. It begins with a single stem at the bottom, which then splits into three smaller branches and each of those separates out again.

The Chiza Geoglyphs (Fig. 11.7) are a series of anthropomorphic and zoomorphic figures created in the Cuesta de Chiza, Tarapacá Region, Chile. They are sited in the middle of the Atacama Desert 80 km south of Arica. They occur on the side of a mountain, and can be seen both from the air and from the ground. They were created using black rocks placed on the sand, forming figures of humans, birds, and other animals. It is presumed that they were created between the years 500 and 900 AD. Discussion of these and other geoglyphs in the area is given by Briones (2006). The distribution of geoglyphs in northern Chile has been associated with pre-Hispanic routes connecting different localities on the coast with oases and inland valleys (Francisco et al. 2018).

Fig. 11.6 Candelabra of Paracas, Peru. 13°47′40.54"S, 76°18′30.70"W. Scale bar = 60 m, ©Google Earth

Fig. 11.7 Chiza Geoglyphs, Chile. 19°12′13.16"S, 70° 0′30.03"W. Scale bar = 70 m, ©Google Earth

11.5 Amazonia

One of the most extraordinary findings of recent decades is that the Amazonian rainforest is not the pristine wilderness that it was once thought to be. There is an increasing body of evidence that its pre-Columbian inhabitants cleared and occupied large tracts. This evidence includes humanly made soils (anthrosols), containing ash, charcoal, and ceramics, and many geoglyphs. Forest clearance for cattle ranching and agriculture has revealed the presence of geometric forms under areas that had previously been overlain by dense forest, and these features have been identified and mapped using GE (Ranzi et al. 2007; Pärssinen et al. 2009), drones, and LiDAR (Fig. 11.8). They are especially well developed in the southwestern region of the western Amazon basin, and most predominantly in the eastern portion of the state of Acre. Together with similar discoveries in the Lower Amazon, Central Amazonia, Xingu, and Mojos and Baures in Bolivia, the Acre geoglyphs have radically changed our understanding of the long-term human impact on the Amazonian rainforest. To date, more than 450 prehistoric earthwork sites have been identified in the state of Acre alone. They are contiguous ditch and bank structures (averaging around 11 m wide and 3 m deep) that enclose typically circular and quadrangular-shaped areas of between 3 and 10 ha in extent (Watling et al. 2018). They are thought to have been made by indigenous groups that inhabited the region between approximately 200 BC and 1300 AD and that they were collective social spaces for ceremonial, symbolic, ritualistic, and residential use (Pärssinen et al. 2020). They have been placed on the World Heritage Tentative site list (https://whc.unesco.org/en/tentativelists/5999/) (Accessed 4th April 2025).

Fig. 11.8 Geoglyphs in cleared regions in the Acre region of Brazil. Top: 10°35′42.78"S, 67°41′15.95"W. Scale bar = 200 m, ©Google Earth. Bottom: 10°35′50.64"S, 67°48′9.60"W. Scale bar = 200 m, ©Google Earth

11.6 North America

Serpent Mound (Fig. 11.9) is a National Historic Landmark built by the ancient American Indian cultures of Ohio, USA. It is an effigy mound (a mound in the shape of an animal) representing a snake with a curled tail. The Serpent Mound extends over 419 m in length, and is 30–100 cm in height. It includes an oval embankment at one end, which has been interpreted variously as the serpent's eye, part of its head, or a secondary object, such as an egg, grasped in the serpent's open jaws. Its date is subject to debate, but it may have been built by the Adena culture at around 300 BC, but others contend it was built by the Fort Ancient culture around 1100 AD (Lepper et al. 2018). Perhaps it was built in two stages (Herrmann et al. 2014). It is a World Heritage Site (https://whc.unesco.org/en/tentativelists/5248/) (Accessed 4th April 2025). Also in Ohio is the Hopewell Ceremonial Site (https://whc.unesco.org/en/list/1689/) (Accessed 4th April 2025) (39°22′35.09"N, 83° 0′15.51"W). This has many mounds and enclosure complexes and may date back to 2000–1600 years ago.

In the Colorado Desert of California, USA, are the Blythe Intaglios (Fig. 11.10) (Hill 2015). These contain three human figures, two four-legged animals, and a spiral. The largest human figure in the Blythe Intaglio group is 52 m long. They were created by scraping away surface layers of darker rocks or pebbles to reveal an underlying lighter-coloured sediment. Many of the intaglios may date from the prehistoric period, but their age and the identity of their creators are still uncertain. It is commonly stated that members of the Mohave and Quechan tribes believe that the human figures represent Mastamho, a creator god, while the animal figures represent Hatakulva, a mountain lion who helped in the act of creation.

Fig. 11.9 Serpent Mound, Ohio, USA. 39° 1′32.93"N, 83°25′48.48"W. Scale bar = 60 m, ©Google Earth

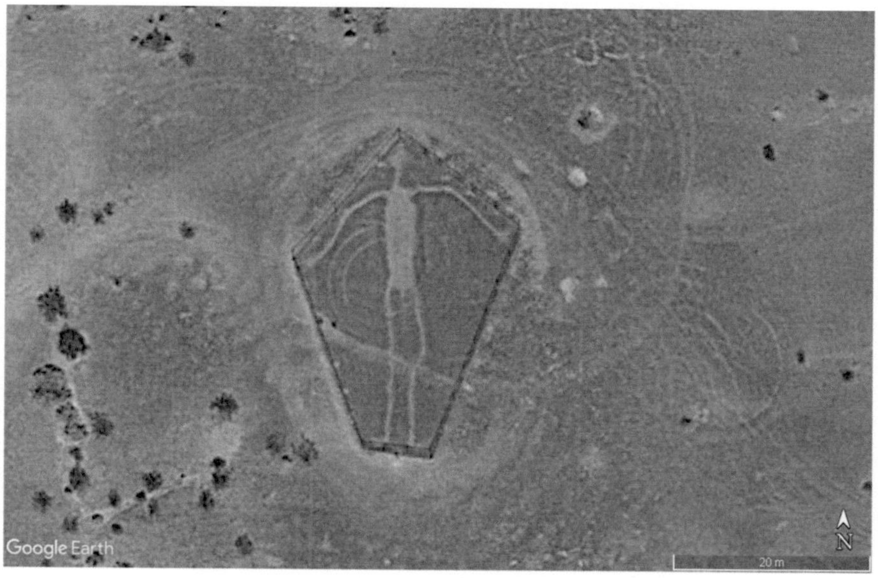

Fig. 11.10 Blythe Intaglios, California, USA. 33°48′2.13"N, 114°31′55.23"W. Scale bar = 20 m, ©Google Earth

11.7 Kazakhstan

One final group of geoglyphs that have been described recently are those of the Turgai region of Kazakhstan (Matuzeviciute et al. 2016) (Fig. 11.11). They date back to c 800 BC but their function is unknown. To date, around 60 geometrical earthworks in total have been identified, concentrated in the southern territory of the Turgai steppe. They vary in size, measuring from 90 to 436 m in length along their major axis. They are laid out in various forms, including a square with diagonal lines, crosses, circles, simple rows or lines, and, in one case, a swastika. The Ushtogaiskii geoglyph is built of 101 individual earthen mounds. The diameter of each of these is 10–12 m, and they are c 1 m in height. The length of each side of the square is 287 m, and that of the internal diagonal lines is 406 m. The Turgai Triquetra (swastika) is a mound which is shaped like a three-armed swastika with a diameter of 94 m.

11.8 Epilogue

In this book I have attempted to show that GE, which has increasingly good resolution, is a valuable resource for visualising landscapes. It also can show how they have changed in recent decades, and by comparison with other sources, such as traditional air photographs, can reveal how they have altered over longer time spans of a century or so. As we saw in Chap. 1, this has been especially valuable in demonstrating the ravages that have been inflicted on some landscapes as a result of war and ideological fanaticism, or because of other pressures such as urbanisation. I have attempted to take a truly international perspective, and to treat a wide range of phenomena dating back to the start of the Holocene. I have deliberately not dealt with some sites and phenomena with which most people are highly familiar. Equally, because of their importance, I have included many UNESCO World Heritage Sites. Also, because they are of limited archaeological interest, I have not included

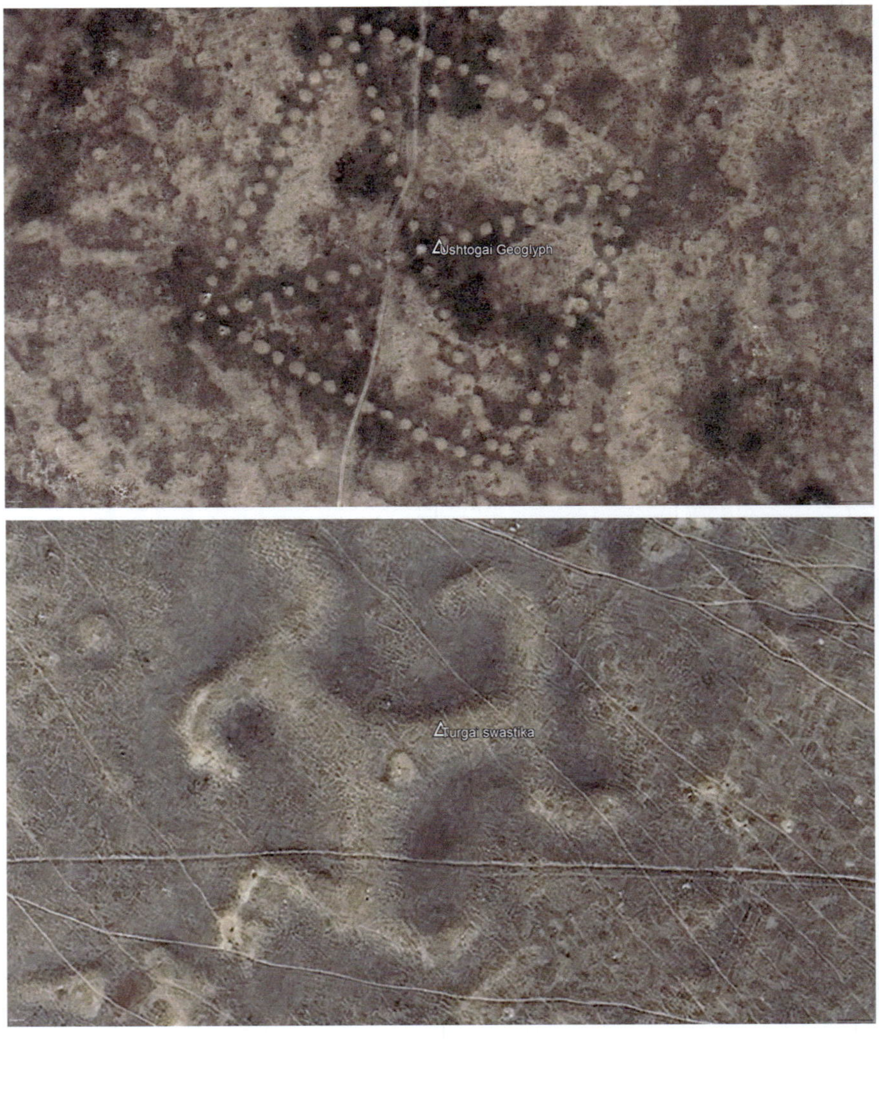

Fig. 11.11 Steppe Geoglyphs of Kazakhstan. Top: The Ushtogaiskii Geoglyph, Kazakhstan. 50°49′58.57"N, 65°19′35.07"E. Scale bar = 200 m, ©Google Earth. Bottom: The Turgai swastika. 50° 6′10.15"N, 65°21′41.31"E. Scale bar = 50 m, ©Google Earth

many modern sites, such as industrial plants or transport links. I have tried to go for the intriguing and the aesthetically beautiful, and for sites which archaeologists and historical geographers have already studied. I have also tried, as much as possible, to include sites for which I have ground images. Alas, I have been unable to include certain sites because they are too small in scale to be clearly shown on current GE imagery. Nonetheless, I hope all the images used here will enthuse and intrigue researchers, students, and travellers alike.

References

Abram DR (2022) Aerial atlas of ancient Britain. Thames and Hudson, London, p 272

Barge O, et al. (2016) The 'desert kites' of the Ustyurt plateau. Quatern Int 395:113–132

Betts A, Burke D (2015) Desert kites in Jordan–a new appraisal. Arab Archaeol Epigr 26(2):74–94

Briones L (2006) The geoglyphs of the north Chilean desert: an archaeological and artistic perspective. Antiquity 80:9–24

Crassard R et al (2015) Addressing the desert kites phenomenon and its global range through a multi-proxy approach. J Archaeol Method Theory 22:1093–1121

Francisco GI, Gloria CB, Gonzalo PG (2018) Signs in the desert: geoglyphs as cultural system and ideology (northern Chile). In: Troncoso A, Armstrong F, Nash G (eds) Archaeologies of rock art. Routledge, London, pp 130–149

Giannelli G, Maestrucci F (2018) Desert kites in the Libyan Sahara: new evidence from remotely sensed images. Antiquity 92:e3

Giannelli G, Maestrucci F (2019) Desert kites in the Tripolitania region: new evidence from satellite imagery. Antiquity 93:e26

Groucutt HS, Carleton WC (2021) Mass-kill hunting and Late quaternary ecology: new insights into the 'desert kite' phenomenon in Arabia. J Archaeol Sci 37:102995

Helms SW, Betts A (1987) The desert 'kites' of the Badiyat esh-Sham and northern Arabia. Paléorient 13:41–67

Herrmann EW et al (2014) A new multistage construction chronology for the Great Serpent Mound, USA. J Archaeol Sci 50:117–125

Hill B (2015) Blythe Intaglios: the impressive anthropomorphic geoglyphs of the colorado desert. Ancient Origins (https://www.ancient-origins.net/searchall/Blythe%20Intaglios) (Accessed 4th April 2025)

Jones D (2020) The inca empire—an illustrated history. Lorenz Books, Dayton, Ohio, p 512

Kempe S, Al-Malabeh A (2013) Desert kites in Jordan and Saudi Arabia: structure, statistics and function, a Google Earth study. Quatern Int 297:126–146

Kennedy D, Bewley R (2004) Ancient Jordan from the air. British Academy, London, p 282

Lepper BT, Duncan JR, Diaz-Granádos C, Frolking TA (2018) Arguments for the age of serpent mound. Cambridge Archaeol J 28(3):433–450

Lombard M, Lotter MG, Van der Walt J, Caruana MV (2021) The keimoes kite landscape of the trans-gariep, South Africa. Archaeol Anthropol Sci 13(6):90

Malkinson D, Bar-Oz G, Gasparyan B, Nachmias A, Gershtein EC, Nadel D (2018) Seasonal use of corrals and game traps (desert kites) in Armenia. Quatern Int 10:285–304

Matuzeviciute GM, Logvin AV, Shevnina I, Seitov AM, Feng J, Zhou L (2016) OSL dates for the ancient geometric earthworks of Kazakhstan. Archaeol Res Asia 7:1–9

Miles D, Palmer S, Lock G, Gosden C, Cromarty AM (2014) Uffington white horse and its landscape: investigations at white horse hill, Uffington, 1989–95, and Tower Hill, Ashbury, 1993–4. Archaeol J 161(1):243–244

Morcom T, Gittos H (2024) The cerne giant in its early medieval context. Speculum 99(1):1–38

Oetheimer C, Oetheimer Y (2021) New enigmatic geoglyphs in the Indian thar desert: The largest graphic realizations of mankind? Archaeol Res Asia 27:100290

Papworth M (2021) Cerne abbas giant: four excavation trenches and interim report of OSL dating. Proc Dorset Nat Hist Archaeol Soc 142:239–240

Pärssinen M, Schaan D, Ranzi A (2009) Pre-columbian geometric earthworks in the upper Purús: a complex society in western Amazonia. Antiquity 83:1084–1095

Pärssinen M, Balée W, Ranzi A, Barbosa A (2020) The geoglyph sites of Acre, Brazil: 10 000-year-old land-use practices and climate change in Amazonia. Antiquity 94:1538–1556

Pollard J (2017) The Uffington White Horse geoglyph as sun-horse. Antiquity 91:406–420

Ranzi A, Feres R, Brown F (2007) Internet software programs aid in search for Amazonian geoglyphs. Eos 18(21):226–229

Repper R et al (2022) Kites of Al Ula county and the ḥarrat 'Uwayriḍ Saudi Arabia. Arab Archaeol Epigr 33(1):3–22

Rollefson GO, Athanassas CD, Rowan YM, Wasse AM (2016) First chronometric results for 'works of the old men': late prehistoric 'wheels' near Wisad Pools, Black Desert, Jordan. Antiquity 90:939–952

Sakai M et al (2024) AI-accelerated Nazca survey nearly doubles the number of known figurative geoglyphs and sheds light on their purpose. Proc Nat Acad SCi 121(40):e2407652121

Storemyr P (2011) The ancient stone-built game traps at Gharb Aswan and beyond, Lower Nubia and Upper Egypt. Sahara 22:15–28

Sukumaran P, Garge T, Mane MM, Chauhan PR (2025) Geospatial context of the Ukshi geoglyph in the Konkan zone of Maharashtra, western India. Anthropologie 13:103349

Svizzero S, Tisdell C (2018) Desert kites: were they used for hunting or for herding? A review of the recent academic literature. J Zool Res 2(4):7–28

Valenzuela D, Clarkson PB (2020) Geoglyphs. In: Smith C (ed) Encyclopedia of global archaeology. Springer, Cham, pp 4538–4551

Watling J, Mayle FE, Schaan D (2018) Historical ecology, human niche construction and landscape in pre-columbian Amazonia: a case study of the geoglyph builders of Acre, Brazil. J Anthropol Archaeol 50:128–139

Index

A. Goudie, *Landscape Heritage of the World with Google Earth*, https://doi.org/10.1007/978-3-031-95555-6

J

K

L

M

N

O

P

MIX
Papier aus verantwortungsvollen Quellen
Paper from responsible sources
FSC® C105338

If you have any concerns about our products,
you can contact us on
ProductSafety@springernature.com

In case Publisher is established outside the EU,
the EU authorized representative is:
Springer Nature Customer Service Center GmbH
Europaplatz 3, 69115 Heidelberg, Germany

Printed by Libri Plureos GmbH
in Hamburg, Germany